Patrick Moore's Practical Astronomy Series

For further volumes:
http://www.springer.com/series/3192

Astronomy with a Budget Telescope

An Introduction to Practical Observing

Second Edition

Patrick Moore
John Watson

 Springer

Patrick Moore
Selsey, UK

John Watson
Old Basing, UK

ISSN 1431-9756
ISBN 978-1-4614-2160-3   e-ISBN 978-1-4614-2161-0
DOI 10.1007/978-1-4614-2161-0
Springer New York Dordrecht Heidelberg London

Library of Congress Control Number: 2011944427

© Springer Science+Business Media, LLC 2012
All rights reserved. This work may not be translated or copied in whole or in part without the written permission of the publisher (Springer Science+Business Media, LLC, 233 Spring Street, New York, NY 10013, USA), except for brief excerpts in connection with reviews or scholarly analysis. Use in connection with any form of information storage and retrieval, electronic adaptation, computer software, or by similar or dissimilar methodology now known or hereafter developed is forbidden.
The use in this publication of trade names, trademarks, service marks, and similar terms, even if they are not identified as such, is not to be taken as an expression of opinion as to whether or not they are subject to proprietary rights.

Printed on acid-free paper

Springer is part of Springer Science+Business Media (www.springer.com)

Preface
to the Second Edition

It's been eight years since we finished the first edition of this book. In 2011, Patrick's uninterrupted record-breaking run of his BBC television program, *The Sky at Night,* passed its 700th episode and was celebrated with a TV special looking back on its long, long history. In the past decade, other important things have happened in the amateur astronomy world.

First, the technology of telescopes and their mounting has improved, along with the value you get for your money. Everything seems to be made in China. Computer-controlled ('go-to') telescopes are commonplace and less expensive than they were, and telescope optics are in general very good or excellent, even in budget telescopes.

The second thing is the rise of the Internet as a vehicle for shopping as well as for information. This has had a detrimental effect on Main Street (or High Street in the UK) shops in all areas, and that in turn means that it is harder than it used to be to find a decent astronomical telescope in any non-specialist shop.

Fortunately, shopping for a telescope on the Internet is easy, *provided you know what you are looking for.* The product knowledge of the specialists will be far better than you would get in a department store, and the after sales service will be better informed (and if you are lucky, actually better). The best advice for this decade is this: before buying, do your research thoroughly using the Internet as a tool, and looking at any user-feedback (not from the retailers' web sites – they'll only put in the good reports!) you can discover. Then buy from a specialist in astronomical equipment.

Third, digital photography has become the norm, and beautiful classic 'wet film' cameras have been consigned in their thousands to, at best, attics and sales of collectibles, or at worst, recycling. The reason is that digital is simply better. Much better. There is no film to buy, processing costs are almost zero, you don't have to wait to see your results, and it is simple to carry out basic image-enhancement on a PC to improve astronomical images beyond recognition. Advanced techniques such as image 'stacking' offer even better results, and there is no longer any need to sit at the telescope and guide it for many minutes to get an acceptable image of a faint object. You can either stack lots of short exposures, or for a little more money, buy a telescope that will do the guiding for you automatically.

So what's new in this revised edition of *Astronomy with a Budget Telescope*?

Some of the telescopes for which we did reviews are now unavailable and others have taken their place, so there are some different reviews. We've added some advice about buying used astronomical equipment (especially telescopes), either face-to-face or on the Internet. We've put in more about using digital cameras and webcams for astronomical photography. We've extended the section on observing the Sun – the recommended techniques for safe observing have changed a little – and we have even talked a bit about the least expensive of the new generation of hydrogen-alpha solar telescopes.

As to the stars, they haven't changed very much! The new Moon maps reflect technical changes rather than anything that's happened on the Moon, Saturn's rings are opening out for a more spectacular view of the sixth planet, and the two components of the binary star Arich have obligingly moved round their orbits to provide us with an increasingly easy view of them.

We hope you enjoy this book and find it helpful, and wish you the traditional astronomers' greeting: 'Clear skies!'

Sir Patrick Moore, CBE, FRS
John Watson, FRAS

Contents

Chapter 1

Introduction

Fig. 1.1 Sir Patrick Moore and John Watson in Patrick's garden at Selsey, England

P. Moore and J. Watson, *Astronomy with a Budget Telescope: An Introduction to Practical Observing*, Patrick Moore's Practical Astronomy Series, DOI 10.1007/978-1-4614-2161-0_1, © Springer Science+Business Media, LLC 2012

"If I want to make astronomy my hobby, what sort of telescope do I need?"

This is a question which has been asked countless times, and there is not always a straightforward answer, because so much depends upon the interests and the circumstances of the beginner. However, there are various points to be borne in mind at once, and these are important, because it is only too easy to make a wrong choice and be bitterly disappointed.

Astronomy can become a hobby with no optical equipment at all. There is a great deal to be seen with the naked eye, and it is even possible to carry out really useful scientific observations. The very first step is to master the essentials, and then learn your way around the night sky – which is by no means difficult, because the stars do not move appreciably with respect to each other; it is only our near neighbors, the members of the Solar System, which wander around. The constellation patterns we see now are to all intents and purposes the same as those which must have been seen by King Canute, Julius Caesar, or the builders of the Pyramids. The trick is to identify a few unmistakable groups, and then use these as guides to the rest.

The first step might be to obtain a pair of binoculars. These have many of the advantages of a small telescope and they can be relatively inexpensive. And if you decide astronomy is not for you, you can always use them for other things. Binoculars are classified according to the diameters of the objective lenses – that's the big lenses at the front, which govern the light-gathering power. Objective lens diameters are always given in millimeters. Thus "7×50" means a magnification of 7, with each objective 50 mm (1.97 inches) in diameter. It is probably wise to keep to a magnification of no more than 12. A higher power often means a larger pair of lenses, and the binoculars become too heavy to be comfortably hand-held. Much more important, they will also be *very* difficult to hold steady, so that some sort of mounting is necessary. We have seen recently advertisements for little pocket-sized binoculars with "massive 50× magnification" – useless for astronomy and extremely difficult to use for anything else. Even proper binoculars, with large-diameter lenses and good optics, are expensive and for astronomical use can never match an astronomical telescope – price for price – for magnification, or even light grasp.

So like many amateur astronomers these days, you may want to begin by buying a telescope.

The Good, the Bad and the Ugly

The Good

The very best amateur telescopes tend to be sold by a few specialist shops and suppliers, who advertise mainly in the Astronomy magazines. The top manufacturers of commercial telescopes are mostly American, although the actual manufacturing is generally done in China. These telescopes represent, for the most part, great value. Many are for specialists who know what they are doing, or at least who know what they want to do: serious amateurs. Such telescopes feature superb optics, computer control and a whole range of standard accessories. That said, they are expensive. Good value, but expensive.

Not everyone can afford the best and most expensive telescope, and indeed not everyone wants to afford the best – not right away at any rate. The alternative is a visit to the "catalog store", main-street camera/video/binoculars/TV/telescope shop, or purchase by mail over the Internet.

Not so many years ago, to buy such a telescope would have been foolhardy. This is no longer the case. The best of the telescopes sold through such outlets are now quite respectable. They can be used for real astronomy, and are of good quality, and we will talk about experiences with two of the best ones in this book. Many are excellent, almost incredible, value. You usually get a lot for your money.

Unfortunately, that doesn't apply to all of them, which is the reason we put this book together, in the hope that it will help you in choosing and using a low-cost astronomical telescope.

The Bad

Fig. 1.2 Not quite what it seems – a small refractor fitted with an aperture stop

One problem is that a bad telescope does not necessarily betray itself by its appearance, and there still are some very bad telescopes on the market. A typical example is shown in Figure 1.2. It was bought by one of us (Patrick) at a charity sale, for the sum of £1 (about $1.60 at the time or writing), but the initial selling price was, we understand, £35 (almost $60!). It is everything that a telescope should *not* be!

The objective lens is 2 inches (50 mm) across, but a cunning 'stop' – like a big circular washer – inside the tube cuts this down to no more than an inch. That means that this telescope really has a tiny objective lens – the stop is there because the rest of that 'big' 2-inch lens is of such poor quality it would actually detract from the image. The little tripod mounting is about as steady as a blancmange (or Jell-o, if you've never experienced blancmange), and sighting any celestial object is more or less impossible Figure 1.3.

Wobbly mountings are still regrettably commonplace. The best mountings at the low-cost end of the market at present are tripods of quite stout aluminum construction, with legs that slide in and out and are locked by knobs. 'Telescopic' tripods made of concentric tubes (like older photographic tripods) are often too rickety to use.

Fig. 1.3 An inadequate tripod

On the subject of wobbly mountings, it is important to make sure that the bearing head (that allows the telescope to move) is up to the job. We have both seen nicely engineered bearing heads used on a small reflector very effectively... but also the identical head sold with a 3-inch (80 mm) refractor that is far too long to allow it to be supported properly.

The Ugly

We feel compelled to say something about 'toy' telescopes at this point. It is possible to buy, rather cheaply, quite nice looking toy telescopes for children. Some of these *look* like astronomical telescopes and have names like 'Spacewatcher',[1] although mostly they are actually terrestrial telescopes and not astronomical ones. (When you look through an astronomical telescope, the picture you see is upside-down. This doesn't matter in the least because you can observe a star, planet, or even the Moon just as well upside down. Terrestrial telescopes produce an upright image, but at the expense of optical quality and a wide enough field of view for astronomical use).

Perhaps worse, some of them are sold under names that *sound* as if they might offer good quality instruments – for example, the 'National Geographic', a reputable magazine in the UK, puts its name to some utterly dreadful toy telescopes. If you want to buy a telescope for a youngster of any age, *please* don't buy one of these toys.

Children are immensely enthusiastic about nearly everything new but are easily disgruntled if a new toy doesn't live up to expectations. A toy telescope is made for a 'toy' price, and its skimpy construction, with cheap lenses (often plastic) makes disappointment inevitable. It is designed to look good, but not necessarily to work properly and will actually be much harder to use than a 'real' telescope – surely the very last thing that's needed for children – and the results are unlikely to be acceptable to observers of any age. Either buy something from the grown-ups' part of the catalog, or settle for inexpensive (but *not* toy) binoculars and a suitable junior astronomy book as a first step.

We have seen – although mercifully not recently – 'Sun diagonals' and 'Solar filters' that are meant to be used at the *eyepiece* of a telescope. These are dreadfully dangerous, and with only moderate misuse or misfortune can cause permanent blindness in a second. If you ever come across one – perhaps along with a telescope at a bring and buy sale – make sure it is destroyed. You could be saving someone's sight. The *only* safe solar filters are specialist, and often quite expensive, glass metalized 'full aperture' filters designed to be used over the front of the telescope, and even then you have to be absolutely certain they have not been scratched or otherwise damaged.

A Good Idea for Newcomers

Joining, or at least visiting, an astronomical club or society is always a good move. You will find people who are very willing to give on-the-spot advice. Not everyone wants to do this, or even has an astronomy society nearby – in which case the advice in this book should be doubly useful…

---

[1] Not to be confused with "Sky-Watcher" telescopes, which are rather good. There is a review of one of them in Chapter 8.

Chapter 2

How to Buy a Budget Telescope

We should first reiterate that by 'budget' telescopes, we are in the main referring to telescopes that can be bought through non-specialist shops, department stores, and on the Internet.

We mean those telescopes that are not manufactured by one of the 'top' astronomical equipment suppliers (including, but not limited to, Meade, Celestron, TeleVue, Takahashi and Orion). We mean those telescopes that are offered for sale for less than about £250 ($400).

Units

Before going any further, let us clear up the question of units of measurement. Frankly, the Imperial system is particularly convenient and is preferred by many people, but in Britain and increasingly in the US, telescope dimensions are often given in SI (metric) units – and not uncommonly in a mixture of the two! – so here are some useful conversions:

Inches	Centimetres	Millimetres
(rounded to one Decimal place)		
1	2.5	25.4
1½	3.8	38.1
2	5.1	50.8
2½	6.3	63.5
3	7.6	76.2

(continued)

P. Moore and J. Watson, *Astronomy with a Budget Telescope: An Introduction to Practical Observing*, Patrick Moore's Practical Astronomy Series, DOI 10.1007/978-1-4614-2161-0_2, © Springer Science+Business Media, LLC 2012

Inches	Centimetres	Millimetres
3½	8.9	88.9
4	10.2	101.6
4½	11.4	114.3
5	12.7	127.0

Thus a 60-mm telescope – a favourite of advertisers – will be equivalent to just under 2½ inches (actually 2.4 inches).

Specifications

Do your research: look at the Internet sites or your mail order catalog (or the advertising brochure), and – especially if the telescope isn't from one of the major manufacturers – be on your guard!

Remember, telescopes are of two kinds: *refractors* and *reflectors*. A refractor collects its light by means of a large lens, known as an *objective* or sometimes *object-glass*, while a reflector uses a curved mirror, known as the *primary mirror*, to gather the light. The objective or mirror produces an image of the star, Moon (or whatever you happen to be observing), and you view this image with an *eyepiece*, which is little more than a powerful magnifying glass; this is where the actual magnification takes place. Every telescope should be equipped with at least two eyepieces of different powers.

It is the diameter of the main mirror, or object-glass, which really counts, and it is a fair rule that the *maximum useful magnification* obtainable is 25× per inch of the primary mirror diameter (the diameter is more usually referred to as the mirror's *aperture*), or the same as the aperture of the primary mirror in millimeters. Thus for a 3-inch refractor, the maximum useful power is $3 \times 25 = 75$. (3-inch aperture is about 76 mm. Near enough!). If you use a more powerful eyepiece – that is one with more magnification – the image won't show any more detail unless the optical system of the telescope is pretty much perfect – unlikely in a budget instrument. More probably, higher magnification will result in an image that is so faint or fuzzy that it will be completely useless.

The Art of Copywriting I: Magnification

It is would be unfair to expect the person who writes the selling copy for a general web site, or for mail-order stores' huge catalogs to have perfect product knowledge. That said, at least *some* product knowledge ought to be a requirement!

In 'selling' telescopes, the first thing that seems to come to the copywriters' minds is *magnification*. After all, a telescope makes things look bigger, doesn't it? So by this reasoning, the more magnification the more impressive the telescope will be and the more people will rush to buy it…which means the advertising copy will tend to quote a ridiculously high 'theoretical magnification.'

The 'theoretical magnification' can of course be anything you want, because it is very simply the focal length of the primary lens or mirror divided by the focal length of the eyepiece. (The focal length of a lens, or of a concave mirror, is the distance away from the lens or mirror at which an image is formed).

A telescope with a 40-inch (1 metre) focal length primary lens and a 1/10 inch (2.5 mm) focal length eyepiece has a 'theoretical magnification' of 40/0.1 = 400×. Throw a 2× Barlow lens into the mix (more about that later) and your little telescope has 'theoretical magnification' of a whopping 800×!

That isn't to say you can actually *use* the telescope at this magnification. Remember that the astronomers' rule-of-thumb for the maximum *usable* visual magnification of any telescope is about 25× per inch of aperture. By that reckoning a 4-inch mirror or lens should allow you to use a magnification of 100×.

Increasing the magnification beyond this will simply result in a bigger but dimmer and fuzzier image. You won't see any more detail. Some of the best observing can be done at a low magnification of 30× or so – only three times that of average binoculars – because a telescope's light-gathering and resolving power provides spectacular views of faint objects at low magnification.

To be fair, some (but not all) advertisers also give a figure for 'optimum usage' – meaning magnification – but even this can be too high, maybe double or triple the usable maximum figure. What is optimum depends on what you are looking at, too.

Any telescope advertised by magnification alone is suspect, although it is not necessarily bad – there is every chance that it is only the ignorance of the copywriter that is to blame.

A mail-order catalog advertisement for the *'Bushnell Voyager 675 × Astronomical Reflector Telescope'* provides a fairly typical example. Even allowing for the execrable grammar ('reflector' is a noun; its adjective is 'reflecting'), the advertisement is unhelpful. Nothing is said about the diameter of the mirror, which proved to be 5 inches (125 mm)! There is therefore no conceivable way that this telescope could be used practically with a magnification of '675×' quoted in the copy.

True, the advertisers have kept on the right side of the law by adding 'optimum usage up to 225 times magnification', and if the optics were really good this would be sensible and accurate, but to claim that the telescope is a '675×' instrument is, to put it mildly, well… pushing the limits.

Another favorite trick is to show a picture of the telescope with a large-diameter dew-cap[1] over the end of the tube, making the aperture look wider and the telescope longer. There is no technical reason to make the dew-cap any wider than the rest of the telescope tube.

And, of course, there is that ultimate deception – a stop inside the tube to cut down the aperture and disguise the appalling quality of a lens or mirror.

---

[1] A dew-cap is like a camera's lens hood, but it's there to help keep dew off the lens rather than stray light.

The Art of Copywriting II: Primary Optics

The *Goldline Optical T70* – now out of production – was described in the catalog we bought it from as a 'reflector telescope.' Manifestly it isn't – it has a '60 mm objective lens' (mentioned in the copy after the 'theoretical magnification', of course).

It would be nice to know what *kind* of objective lens it has, but this information is almost always omitted.

A plain, single-element (i.e. one-piece of glass) lens suffers from an optical fault called *chromatic aberration*. Chromatic aberration means that light of only one color is brought to a focus; in practice, the result is rainbows surrounding bright objects and the impossibility of focusing the telescope sharply. It was chromatic aberration that led Sir Isaac Newton to invent and refine the reflecting telescope, in the mistaken belief that there was no way of fixing it.

Almost all telescope lenses are *achromatic* (from the Greek 'no color'). Such lenses focus light of two different colors (i.e., red and green) in the same plane, and this is actually good enough for most purposes. Newton would have been impressed, because he just didn't think of the idea of using two elements (pieces of glass) of different densities to banish the worst of the effects of chromatic aberration! It is usually safe to assume that modern inexpensive astronomical telescopes (apart from toys) have achromatic lenses.

The very best objective lenses are called *apochromatic*, and are designed to bring three different wavelengths of light to a sharp focus in one plane. But in a low-cost telescope? You must be joking.

A reflecting telescope uses a surface-aluminized mirror to produce the image, instead of a lens. A reflecting telescope does not suffer from chromatic aberration at all, apart from a tiny amount that might be introduced by lenses in the eyepiece.

Spherical mirrors have their own problems, the most basic of which is called (unsurprisingly) *spherical aberration*. This causes fuzzy, difficult-to-focus images and is a result of simple geometry – a concave spherical surface just isn't the right shape to focus incoming light in a single plane. This can be corrected by deepening the spherical curve to make it parabolic. The parabolic shape has to be *exactly* right – a small instrumental error during the manufacture of the Hubble Space Telescope's primary mirror made the whole wonderful $1.5 billion machine more or less useless until it was fixed by a NASA 'repair mission' which installed an optical correcting lens.

Because of the higher cost of a parabolic mirror, you will see in the advertising copy for low-cost telescope phrases like, '*Genuine 100 mm spherical aluminized mirror*', as if the 'spherical' bit is something good! It isn't. 'Parabolic' or 'paraboloid' is ideally what is needed. A spherical mirror will suffer from spherical aberration and will not produce the sharpest images. That said, we're not looking for HST quality, and a small aperture (diameter of the primary mirror) relative to its focal length reduces the severity of the problem.

They usually tell you the primary lens or mirror's focal length (remember, that's the distance from the lens or mirror to the point at which the image is formed). You'll need to know this to work out the magnification with any given eyepiece.

The most vital statistic – all else being equal – is the diameter of the primary lens or mirror. One of the most important factors is a telescope's light grasp, for it is this that determines how faint an object you can see. The light grasp increases as the *square* of the diameter. This means that a 100 mm lens will not take in twice the light of a 50 mm lens, it will take in *four* times as much light. Increase this to 110 mm and that's almost *five* times as much. And so on.

The diameter of the primary lens also determines the maximum possible *resolution* of a telescope (that is, it's ability to see fine detail and separate close stars from each other). Once again, the bigger the better. All else being equal, of course.

The Photographer's Art

One shouldn't really expect a photographer to know how to use an astronomical telescope. However...many of the better budget telescopes have slow-motion controls (more about that later) which allow you to move the telescope slowly to and fro when trying to aim at some astronomical object.

Such slow-motion controls are very often fitted with short flexible shafts. The shaft is bendy, and has a knob on the end. Fairly obviously, you turn the knob to move the telescope. There are always two such controls, and because (also obviously) you will be looking through the telescope's eyepiece when you are using it, it is convenient to situate the knobs close to the eyepiece so you can make adjustments – one of the reasons for the flexible shafts.

Unfortunately the telescope's design usually allows assembly of the flexible shafts in different positions. A lot of harmless fun can be had in browsing the web sites and catalogs to see which photographer has managed to assemble the controls in the *least* convenient position! Getting the knob somewhere down at the bottom of a reflector seems to be a favorite. Pity the eyepiece is near the top...

I (John) even saw a catalog photograph of one reflecting telescope set up so that the *eyepiece* was near the bottom, with the immaculately-groomed photographer's model peering through it in the hope of seeing, presumably, his own feet at 675 × theoretical magnification!

The Art of Copywriting III: Diagonal Prisms

If you are pointing a refractor upwards, the position of the eyepiece often requires that you adopt some bizarre horizontal posture to look through it. A diagonal optical system (sometimes called a 'star diagonal') that bends the light coming to the eyepiece through 90° solves this difficulty, and enables you to get the eyepiece into a reasonable angle for observing.

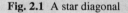
Fig. 2.1 A star diagonal

Light can be turned through 90° by a simple right-angled prism (as used in prismatic binoculars), or of course by a flat mirror. Prisms are better because they don't need an aluminum coating like the mirror does. A mirror is okay – and it's cheaper and lighter than a prism – but over time the coating can get corroded and battered, which of course doesn't happen to a prism.

Well, contrary to what the copywriters say about 'diagonal prisms' being included in the set with a cheap telescope, we haven't been able to find one. They were all flat mirrors. Every one of them.

Accessories

Even low-cost telescopes often come with a formidable array of accessories.

Eyepieces

At least two, and sometimes three, eyepieces are included. You'll probably find the *least* powerful one (the one with the longest focal length) is far and away the most useful. In most of the telescopes we looked at, the most powerful eyepiece provided such high magnification as to be quite useless.

Fig. 2.2 Eyepieces for a budget telescope

Barlow Lens

This mysterious item is often included, and is sometimes described with a magnification all of its own, such as '2× Barlow Lens.' A Barlow lens is simply a negative lens (i.e., things look smaller through it) that makes the focal length of the primary mirror or lens appear longer than it really is. It fits between the eyepiece and the eyepiece holder.

Fig. 2.3 A Barlow lens

The magnification factor marked on the Barlow lens just multiplies the basic magnification. Here's an example. If a telescope has a primary mirror of 48 inch (1,200 mm) focal length and an eyepiece of ½ inch (25 mm) focal length, the magnification will be 48/½ = 96×. With a 2× Barlow lens, this simply becomes 2 × 96 = 192× magnification.

Rack-and-Pinion Focusing

This is simple enough. It means you focus the telescope by operating one or more knobs set at right angles to the eyepiece. The knobs turn a gear and the gear engages a rack (a sort of unrolled gear) that moves the eyepiece in and out and focuses the telescope.

Fig. 2.4 Rack and pinion focusing on a small refractor

The alternative is usually a 'drawtube' which means you just pull the eyepiece in and out to focus it. That's less satisfactory, but is now seldom seen, even on 'toy' telescopes.

Finderscope

Astronomical telescopes have a narrow field of view, which makes it surprisingly tricky to point the telescope at whatever it is you want to look at. To make life easier, a *finderscope* is fitted. This is simply a small low-powered telescope that acts like a 'gun sight' to help you point the main instrument. Finderscopes are 'terrestrial' telescopes in that the image is the right way up, which makes them easier to use as sights. They usually have cross-hairs.

Fig. 2.5 A typical finderscope. Note the three adjustment screws

The finderscope has to be aligned with the main telescope in order to be of any use, and the most common way is for the finderscope to be supported by three adjustable screws that enable you to move it. This is a fiddly task, and it's surprising that the archaic three-screw arrangement is still used, even on expensive telescopes – there are easier ways!

Finderscopes on cheap telescopes are mostly good enough... there's not much more to be said, really.

At the 'top end' of the range of budget telescopes, *red dot finders* are becoming popular. They project a small red spot of light from a red light-emitting diode (LED) onto the field of view, so that you can see the dot against a natural (non-magnified) view of the sky. You then aim the telescope so that the red dot is superimposed over what you want to aim at. Red dot finders are easier to use, and quite accurate enough. The only downside is that they generally use a small battery to power the LED, so you need to remember to turn them off when you've finished using them.

Fig. 2.6 A red dot finder, fitted to a 5-inch reflector

Mountings

You can't hand-hold an astronomical telescope. It's too heavy, and you could never hold it still enough to see anything, even at 'low' magnification. At the very least, you have to mount it on a tripod or on a 'Dobsonian' mounting. All the inexpensive telescopes we tried had tripods supplied with them.

There are three ways to mount a telescope on a tripod.

Ball and Socket

The first and worst way to attach the telescope to a tripod is to use some kind of a ball-and-socket joint, just like you would for a camera. This is a pretty horrible idea, but it's cheap. On the downside, it's also wobbly and almost impossible to manage. To be avoided like the plague.

Fig. 2.7 Ball-and-socket – the very worst mounting for an astronomical telescope!

Altazimuth Mounting

This is a mounting that enables a telescope to be moved in two planes: up and down (**alt**itude) and from side to side (**az**imuth). Most lower-cost telescopes tend to use this kind of mounting.

The first requirement of any telescope mounting is that it must be very, very solid. Anyone who has ever tried to hold binoculars still enough to see the stars will know that the slightest tremble is magnified just as much as the object you are trying to watch. A telescope, even used at a sensible power, will have a great deal more magnification than binoculars and the effects of the slightest shaking are dramatic.

Years ago, the mountings of inexpensive telescopes were utterly dreadful. Most of them would shake in the slightest breeze, take ages to settle down after you had touched the focusing control, and generally make life for the observer thoroughly miserable. Patrick's advice then was, 'If you can only afford a cheap telescope like this, buy a good pair of binoculars instead.' Much to our pleasurable surprise, the situation has – generally – changed for the better.

Most of the small telescopes we tested were mounted fairly rigidly. Some of them rather astonished us – they were excellent in this respect. Typical of these altazimuth mountings is a die-cast metal fork, angled backwards so that the telescope can be pointed more or less vertically upwards if necessary. The telescope tube is supported in the fork by two substantial bolts that can be tightened to lock the tube in altitude (up and down).

The fork is set in the base of the mounting (which is the same as the top part of the tripod) so that it can be rotated in azimuth (horizontally). There is a knob that you can tighten to lock it in place.

Fig. 2.8 A typical altazimuth mounting for a budget telescope

There are two problems relating to basic altazimuth mountings, however solidly they are made.

The main one is the fact that the Earth rotates in space.

The Earth rotates relative to the stars at the rate of roughly one revolution per day, in a counter-clockwise direction looking down on the North Pole. That doesn't sound very fast, but remember that your telescope magnifies a rather small area of sky. Even at its lowest power, your telescope is unlikely to have a field of view of more than ½°. That means you can center a star in the middle of your field of view, and if it happens to be near the horizon it will have disappeared out of sight in a little over one minute! The need for constant re-aiming of the telescope is a major disadvantage to observing.

The second problem relating to the lowest-cost altazimuth mountings is the usual lack of any slow-motion controls.

A slow-motion control is gear-driven, and allows you to move the telescope slowly and smoothly across the sky in any direction. This makes it possible – although not easy: it takes practice to turn the two knobs by just the right amount to follow an object's (apparent) movement through the night sky – to follow celestial objects. Unfortunately such controls cost a little more and aren't generally available on the very cheapest telescope mountings.

Equatorial Mounting

Better telescopes – even in this low price bracket – have *equatorial* mountings.

An equatorial mounting is basically an altazimuth mounting tipped over so that the 'azimuth' axis points towards the north celestial pole. The north celestial pole is fairly close to the bright star Polaris in Ursa Minor (the Little Bear); Southern Hemisphere observers have more of a problem, as the south celestial pole is unmarked by any bright star.

When the 'azimuth' axis is pointed at the pole, it will be parallel to the Earth's axis of rotation. With the telescope aligned on this axis, only *one* movement is needed to follow the stars. Turn the telescope in azimuth (east–west) at the right rate, and it will follow any celestial object. The other control is only needed to correct for misalignments and suchlike.

Aligning the 'azimuth axis' with the celestial pole can be tricky, even with the help of Polaris, the Pole Star, but it gets easier with practice (see Chapter 3). And of course, if the telescope is portable, the axis has to be re-aligned each time the telescope is moved from one place to another.

The 'azimuth' axis, when aligned with the celestial pole, is known as the 'Right Ascension' axis. The reason for this strange name is historical. If you stand next to an equatorial mounting set up in the northern hemisphere facing north, everything on the right-hand side of the north–south bearing axle will rise, or 'ascend', and everything on the left-hand side of the axle will move down towards the ground. The other axis of movement, the one at right-angles to the Right Ascension (R.A. for short) axis is called the 'Declination' axis.

Having an equatorial mounting fitted with slow-motion controls means you can follow a star, planet, or the Moon by turning the telescope in one axis – Right Ascension – only, which is far more convenient than juggling two slow-motion controls that need to be turned at different speeds. It also means that a small electric motor can be used to drive the Right Ascension axis so that – assuming everything is set up properly – the telescope will track a celestial object indefinitely, without any intervention by the observer. And an equatorial mounting is essential if you want to take photographs of the stars and planets – more about that in Chapter 7.

Tipping the mounting over towards the pole produces balance problems. The telescope has to be more or less balanced in any position so as to avoid straining the bearing head or slow-motion gears. And of course, if it were not balanced the telescope would tumble round directly you slackened the locking knobs. The usual design for low-cost equatorial mountings is called a 'German' mounting, and features a heavy counterweight to balance the telescope. Everything has to be carefully balanced so that the telescope is more or less 'neutral.'

Fig. 2.9 A typical equatorial mounting for a budget telescope. Note the setting circles and the slow-motion cable for R.A. The declination cable is not fitted in this picture

An alternative to the German mounting is a 'fork' mounting, in which the telescope is mounted inside a double fork, inclined at the correct angle and supported by a pier or heavy-duty tripod. This is the mounting style favored in the design of most computer-controlled astronomical telescopes. Some models even omit one half of the fork, supporting the telescope on a single strong, and very stiff, upright. A simple built-in computer can easily do the calculations to move the two axes at precisely the rates needed to follow the apparent movement of the sky. This avoids the necessity of having a counterweight and makes for a much more compact unit.[2]

---

[2] An unusual single-support computer-controlled altazimuth mounting was among those we tested – see Chapter 8.

Remember, equatorial mounts always have slow-motion drives, which is a *major* advantage in using the telescope.

Setting Circles

Equatorial mountings usually also have *setting circles*. They can be used to point the telescope at a given object, assuming you know the co-ordinates (Right Ascension and Declination) of the object and the exact time. The setting circles are clearly visible in the photograph of the equatorial mounting above.

Before setting circles can be used, you have to align the mounting quite accurately.

Just *how* accurately you can point your telescope at an astronomical object using setting circles depends on a number of things, including the accuracy of the setting circles themselves. In low-cost telescopes this is unlikely to be better than about 1°. Since this will be more than the telescope's field of view even at low power, it is best to use the setting circles – if you use them at all – for initial alignment (see below) and finding out approximately in which area of sky the object is located. You would need to be very accurate (and rather lucky) to hit any object 'spot on' by using the setting circles of a low-cost telescope. To be blunt about it, we regard setting circles on a budget telescope as more decorative than useful!

Telescopes and Their Mountings: A Thought Before Buying

At first thought, you would imagine the quality and size of a telescope to be overwhelmingly the major factor in deciding which one to buy. It is vital to understand that this is definitely *not* the case: it's only half the story.

The telescope *mounting* is equally important. A reasonable 2-inch (50 mm) telescope on a good equatorial mounting with slow-motion controls is actually far more useful than a good 4-inch (100 mm) telescope on an indifferent mounting, or indeed on *any* mounting that doesn't have slow-motion controls.

It is difficult to overstate just how tricky it is to aim your telescope at celestial objects if the mounting is poor, and no telescope is any use at all unless you can point it at what you want to see! Add to that the fact that without slow-motion controls you will have to realign it at least every few minutes at best to keep an object in the field of view and you'll begin to see that for any kind of serious work (or even for hassle-free astronomical sight-seeing) slow-motion controls and a mounting that won't wobble are pretty well essential.

So overall, what should one look for? Here is our check list, in order of priority.

What to Look For

- A telescope with a fairly large primary lens or mirror. Two-inch (50 mm) is about the smallest that will be of much use.
- A really solid mounting, including an altazimuth or, much better, an equatorial or computer-controlled bearing head.
- Slow-motion controls.
- At least one low power (magnification) eyepiece (much more important than high power).
- A finderscope on an adjustable mount.
- A star diagonal if the telescope is a refractor.
- Rack-and-pinion focusing.

What to Know

- A bargain when you see it – remember that we found the same telescope selling under different brand names, and the more expensive one was almost *twice* the price of the least expensive!
- Your consumer rights – in the UK you are entitled to return faulty goods for a refund (even if you can't prove they were faulty at the time you bought them) within six months. You are also entitled to claim at least a proportion of the cost of rectifying faults for six years from the date of purchase. If it goes wrong within two years don't be fobbed off with, 'Sorry, it's over a year old and so out of guarantee' – EU law and the UK's *Sale of Goods Act* almost certainly protects you. In the USA the *Uniform Commercial Code* is applicable in most states. All goods carry with them an implied warranty of merchantability and fitness for purpose unless you sign these rights away. You can return faulty products within what they call 'a commercially reasonable period of time.'

What to Avoid

- 'Toy' telescopes.
- Ball-and-socket heads.
- Any telescope with an aperture stop.
- Terrestrial telescopes masquerading as astronomical telescopes (they show upright images).

What to Ignore

- The descriptive advertising copy – try to look for yourself if you can.
- Claims for very high magnification.

Buying a Used Telescope

There isn't a great deal to go wrong with an astronomical telescope, so buying a used one can be a great money-saver. Quite a lot of would-be astronomers put their telescopes into a closet or attic after a few weeks and never get them out again until they decide it was all a bad idea and they'd rather have something else.

So what can go wrong? Well, it's important not to buy something you haven't been able to look at. This is, after all, a scientific instrument and needs to be in perfect, or near-perfect condition. Look at the optical system from the front of the tube. Any serious scratches on an objective lens – or worse, scuffs on a primary mirror – should lead to rejection. Remove the eyepiece and look in that way too: it will show up any damage to secondary mirrors.

Dust doesn't matter too much unless it's severe, as it won't affect the image. Likewise, any minor imperfections in the lenses or mirrors can be ignored.

Some people advocate shining a flashlight down the telescope as a check, but this is really unnecessary as it will always show up dust and other specks, even in a brand new instrument.

Check the mounting. Play in either axis should lead to immediate rejection. A little backlash in the slow motion controls doesn't matter and is actually quite usual. But is the tripod wobbly? If it is, it will probably be very difficult to fix – but maybe it would be worth negotiating a lower price that will allow you to buy a replacement.

Finally, it's hard – but ask for a right to return it after you've tried it out on the stars, in case there's something unobvious wrong with it. If the person selling it is honest and knows it's okay, he or she might well agree to a deal that depends on using it on the first available clear night as a test.

Astronomy: An Expensive Pastime?

If you want a really good telescope, then it will cost money, but let us keep a sense of perspective. You want to travel from London to Edinburgh by rail. You can of course brave the discomfort and overcrowding of an 'economy-class' ticket at about £110 ($175). If you want a reasonably civilized journey, you must travel first-class: £200 ($320).

Those living in the US, or other countries where the delights of the UK rail service are unavailable might like to consider the cost of, say, three auto tires. Or perhaps the cost of two people going to a Broadway show and following up with dinner for two... there won't be a lot of change from $300.

Compare this with the cost of some of the telescopes we talk about here, and you will have to agree that astronomy is not a really expensive hobby after all!

Chapter 3

How to Use a Budget Telescope

As we have noted, much depends upon your circumstances as an observer, and on your main interests. One great advantage of a budget telescope is that it will certainly be lightweight enough to be portable, so that if you happen to live in a badly light-polluted area you can simply put the telescope in your car and drive into the country. If you are not a car owner and live in the middle of a city, the best solution is to move home – but of course this isn't always practical! There's some good advice for city dwellers in Bob Mizon's book (from Springer), *Light Pollution: Responses and Remedies.*

Care and Maintenance

Astronomical telescopes need little maintenance provided they are looked after.

Lenses should be cleaned seldom, mirrors never.

Clean the lenses by breathing on them and wiping them very carefully and gently with a lens-cleaning cloth or tissue (you can get them at opticians) or a laundered handkerchief. Don't use Kleenex or paper kitchen towels, they might scratch: lenses are made out of very soft glass.

Don't attempt to clean, or even touch, any telescope mirrors. They are surface-aluminized and unless you are an expert you will do more damage than good. Small amounts of dirt won't affect the image. If telescope mirrors get very dirty or degraded, ask someone from your local astronomical club to help.

P. Moore and J. Watson, *Astronomy with a Budget Telescope: An Introduction to Practical Observing*, Patrick Moore's Practical Astronomy Series, DOI 10.1007/978-1-4614-2161-0_3, © Springer Science+Business Media, LLC 2012

Dew

When you bring a telescope into a warm room after you have been using it in the cold outdoors, dew will condense on the surfaces, including the lenses and maybe the mirrors. The very worst thing you can do is to attempt to dry the optical components. Dew consists of almost pure (distilled) water and will eventually evaporate without leaving any trace of residue. Just wait for it to go away.

Mechanical Parts

The mechanical parts of a telescope will need little maintenance, probably no lubrication, and hardly any cleaning. Keep the dust off it. That's about all.

Traveling with Your Telescope

Obviously, great care must be taken in moving the telescope around. It's a fairly fragile piece of equipment. It must be said that refractors have the advantage here; reflectors are much more likely to go out of adjustment. One of the 'down sides' of many inexpensive telescopes is the fact that the tripod stands don't fold up very easily. Ease of use tends to be sacrificed to low cost and solidity. It's quite expensive to make a good tripod – like the Meade ETX tripod illustrated here in Figure 3.1 – that is at once extremely solid and easy to fold up.

Cheaper tripods tend to be less solid, and they also tend to require more nuts and bolts to be removed (and kept safely!) when you want to move them.

If you are driving into the country, remember to take everything with you. There is nothing more infuriating than to drive 10 miles or more, unpack the telescope and then realize that the tripod bolts – or all the eyepieces – are still in the hallway of your home!

Fig. 3.1 A typical good-quality tripod for a small telescope

Setting Up an Equatorial Mounting for Visual Use

If you have an altazimuth mounted telescope, there is no setting up needed. Just place the telescope and tripod somewhere where you have a good view of the sky, and you're off…

If you have an equatorially mounted telescope – which allows you to track celestial objects much more easily – then you will have to set it up correctly before you start your night's observing.

The telescope mounting needs to be pointed towards the north celestial pole (in the northern hemisphere of the world – you need to line up on the south pole if you live in the southern hemisphere). This is not difficult, at least in the north.

Assuming you have already assembled the telescope and mounting and lined up the finderscope with the main telescope, this is how it's done.

1. Go to a location where you can see Polaris (the pole star).

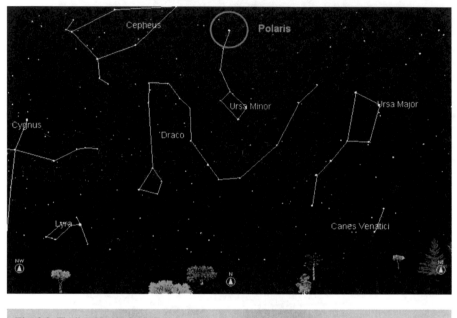

Fig. 3.2 Finding North (in the northern hemisphere)

The chart in Figure 3.2 shows Polaris at the top of the picture, but bear in mind that the stars rotate around the pole, so the sky is unlikely to look the same from where and when you happen to be observing! This chart is looking North, at a latitude of about 50°N. The easiest way to find Polaris is to locate the 'big dipper' – the pan-shape formed by the seven brightest stars in Ursa Major. The two stars opposite the handle were known to sailors as 'the pointers' because they point towards Polaris (and thus North), which is about five times the distance *between* the pointers away from them.

2. Set the telescope's declination axis to exactly 90° and lock it.
3. Now, using *only* the latitude adjustment and the horizontal adjustment on the mounting, aim the telescope exactly at Polaris, using the finderscope or red dot finder to get it roughly right and then a low-power eyepiece: try to get the star right in the middle of the field of view. Once you have done this, lock the latitude and horizontal settings.

You will now be able to track any celestial object (once you have found it) approximately, simply by operating the R.A. slow-motion.

More Accurate Alignment

The above procedure gives an approximate alignment, good enough for most visual observing. If you want to align the telescope more accurately (for example, to prevent the image drifting under high magnification, or if you are attempting astrophotography) then do this:

4. After aligning the telescope as above, make sure that the telescope's declination is set to exactly 90° and locked.
5. While looking through your lowest-power eyepiece, swing the telescope to and fro through at least 180° of R.A. You will notice that Polaris describes a circular path. If it simply whizzes out of the field of view, use a lower power eyepiece.
6. Once again using *only* the latitude and horizontal adjustments on the mounting, make small alterations until Polaris describes the smallest possible circle. If you can't see a complete circle because it won't fit in the field of view, make an estimate using the part of the circle you *can* see. As a last resort, try using the finderscope or red dot finder instead.

There are many other, more accurate methods of setting up an equatorial mounting, but for the telescopes we are discussing in this book, you won't need them.

Computer-Controlled Mountings

Not many computer-controlled mountings fall into the budget category, but you can probably buy a used one inexpensively. Assuming you've bought carefully and know that it's in full working order, just make sure you follow the instructions supplied to set it up accurately. Some telescope computer-control systems, such as Meade's Autostar™, can – like almost all computers – get themselves in a knot. If a 'go-to' telescope persistently refuses to go to your selected objects, then reset the controller and follow the precise, and often quite lengthy, set-up instructions to get it working. User web sites, and books such as Mike Weasner's *Using the Meade ETX: 100 Objects You Can Really See with the Mighty ETX* are invaluable for technical advice.

Chapter 4

Observing the Solar System

The Sun, Moon and planets of our Solar System include some of the most interesting objects for our small telescope.

You cannot, of course, expect to see anything like the views obtained with multi-billion dollar projects like the Hubble Space Telescope. Almost everyone is familiar with at least some of those spectacular images.

But neither should you underestimate the fascination – and wonder – of being able to see details on our nearest astronomical neighbors. Your first sight of ringed Saturn, hanging there like a jewel in the night sky, is something you will never forget!

The Images in These Sections

We have included a high-quality image of each planet, etc., showing a great deal of detail. These images were mostly obtained courtesy of NSSDC and NASA. They show you what you are looking at.

More usefully, we have also included images that show the object as seen visually through a very good budget telescope under clear skies – a kind of benchmark of excellence. This is *not* the same thing as a photograph taken through a small telescope. Astronomical photographs, even those taken with the simplest equipment (see Chapter 7), can show color, and can image objects that are far fainter than you can see through the same telescope.

Remember, the 'reference images' given here relate to the *visual appearance*.

P. Moore and J. Watson, *Astronomy with a Budget Telescope: An Introduction to Practical Observing*, Patrick Moore's Practical Astronomy Series, DOI 10.1007/978-1-4614-2161-0_4, © Springer Science+Business Media, LLC 2012

The Moon

The Moon is dazzling when seen through a telescope, but there is absolutely no risk to your eyesight. A 'lunar filter' is supplied with many telescopes, and is simply a neutral-density filter intended only to reduce the 'dazzle factor' and make the Moon more comfortable to look at.

The Moon is a small world, with a diameter not much more than one quarter that of the Earth, and much less massive. It would take 81 Moons to equal the mass of our world. This means that it has a weak gravitational pull, and has long since lost virtually all its atmosphere, so that if we call it an 'airless world', we are to all intents and purposes correct. It is also waterless and lifeless.

The broad dark plains, easily visible with the naked eye, are still called 'seas', and have been given romantic names such as Mare Serenitatis (the Sea of Serenity), Sinus Roris (the Bay of Dew), Palus Somnii (the Marsh of Sleep) and Lacus Somniorum (the Lake of the Dreamers), but they are bone-dry lava-plains, the site of violent volcanic activity thousands of millions of years ago. There are no clouds on the Moon, so the surface details are always sharp and clear-cut, while the shadows are well-defined and inky black. Local color is absent.

Mountain ranges are common and lofty; many of them form the boundaries of the more regular seas (maria). Thus the Apennines, the most prominent of all the ranges, in part borders the huge Mare Imbrium (Sea of Showers); the peaks go up to well over 15,000 feet (5,000 m). There are vast numbers of isolated mountains, and there are Crack-like features known as rilles or clefts. But the whole scene is dominated by the craters, which are everywhere; they cluster thickly in the highlands, and are also to be found on the seas. They break into each other and distort each other; some are bold and regular, others so ruined that they are barely recognizable. They have been given names of distinguished people, usually astronomers; most of these were allotted in the seventeenth century, though many others have been added since. There are even some unexpected names. Julius Caesar has his crater, not because of his military prowess but because of his association with calendar reform. There are also a couple of Olympians – Atlas and Hercules – while one crater is named Hell. This isn't because of its depth, but because it honors the Hungarian astronomer Maximilian Hell.

Some of the craters are huge, with diameters of well over 150 miles (240 km), while others are tiny pits. In general their walls do not rise to the great heights above the outer country; the floors are sunken, but seen in profile they resemble saucers rather than steep-sided mine-shafts. Many of them have central mountains, or groups of mountains. Thus Theophilus, on the edge of the Mare Nectaris (Sea of Nectar) is 64 miles (102 km) across, with walls rising to 18,000 feet (6,000 m) above a floor upon which is a magnificent, many-peaked central mountain mass. Note, however, that no crater has a central peak rising to a height greater than that of the surrounding rampart. In theory, you could drop a flat lid over Theophilus!

It is possible to see quite a lot of detail on the Moon. This is how Theophilus (the largest of the prominent group of three craters near the bottom centre of the picture) might look on a good night, except that the picture doesn't really convey the dazzling brightness of the Moon's surface.

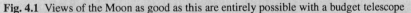

Fig. 4.1 Views of the Moon as good as this are entirely possible with a budget telescope

The lunar craters are impact structures, formed during the 'Great Bombardment', around 4,000 million years ago, when meteorites rained down upon the recently solidified lunar surface. This was followed by widespread volcanism, with magma pouring out from below the crust and flooding the floors of some of the craters, such as 60 mile (100 km) Plato, between the Mare Imbrium and Mare Frigoris (Sea of Cold).

For the last 1,000 million years there have been no major structural developments, so that the dinosaurs must have seen the Moon very much as we see it today.

We have, of course, seen the Moon's surface close up. Here is astronaut Harrison Schmitt of Apollo 17, about as close as you can get to it! He was on his third EVA ('Moon walk') and is next to a large boulder at Taurus-Littrow.

Fig. 4.2 Harrison Schmitt taking a good look at the Moon from very close up!

The Moon completes one orbit in 27.3 days. It spins on its axis in exactly the same time, so that it keeps the same face turned permanently Earthward. (There is no mystery about this apparent coincidence; tidal friction over the ages has been responsible). This means that the visible features are always in the same positions on the disk.

However, the Moon's appearance (which way up it is) depends on what kind of telescope you are using to look at it. The full-Moon pictures on the next pages show three major craters (red labels) and some of the mares.

Fig. 4.3 The Moon as seen with the naked eye or binoculars from the northern hemisphere – North is *up* and West is on the *left*

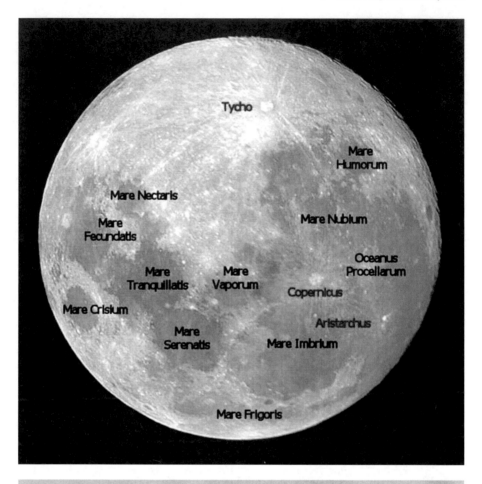

Fig. 4.4 The Moon as seen with an astronomical refractor or Newtonian reflector – South is *up* and East is on the *left*: upside-down compared with the naked-eye view

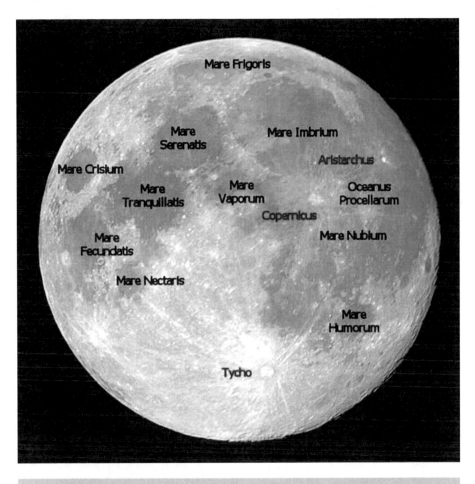

Fig. 4.5 The Moon as seen with a Schmidt-Cassegrain or Maksutov telescope – North is *up* and East is on the *left*: this is a mirror image of the naked-eye view

You will need a lunar atlas for almost any kind of observations of the Moon. Most lunar atlases show South at the top and East on the left. This is upside-down compared to the Moon's visual appearance. The reason is that this is the IAU (International Astronomical Union) standard, corresponding to the observed image seen through a Newtonian reflector or astronomical refractor in the Northern hemisphere. If you are using binoculars, a terrestrial telescope, or just your eyes it isn't too hard to turn the map upside down to find out what features you are looking at.

However, it is very difficult – approaching impossible for most people – to make the mental translation that flips the map to correspond to what we see through a modern commercial Schmidt-Cassegrain or Maksutov style of telescope – that's North up but still East on the left. Our brains just don't take easily to it. Fortunately there are one or two books of lunar maps that are specifically designed for Schmidt-Cassegrain telescopes, for example Cook's *The Hatfield SCT Lunar Atlas*. The 'normal' version of this atlas, *The Hatfield Photographic Lunar Atlas* is of course excellent for users of Newtonians. Both books are published by Springer.

The simple full-Moon maps offer no more than the sketchiest outline of the Moon (they are intended just for orientation), but may serve as a preliminary guide. For example, look at Mare Crisium. Lunar craters are basically circular, but when they lie away from the centre of the disk they are foreshortened into ovals. This effect is well demonstrated by the Mare Crisium, which appears elongated in a north–east sense. Actually, the north–south diameter is 280 miles (450 km), while the east–west diameter is 350 miles (560 km). Figure 4.6 shows how Mare Crisium should look through a modest telescope, in the most favorable conditions:

Fig. 4.6 As good a view as you are likely to see of Mare Crisium with a budget telescope

Until the Space Age we had no direct information about the far side of the Moon, which is always turned away from us – though slight 'rocking' movements (known as 'libration') mean that altogether we can see 59% of the total surface (though of course no more than 50% at any one time). Then in 1959, the Russians sent an unmanned space-craft, Lunik 3, on a round trip, and obtained the first images of the unknown regions. Since then we have mapped the whole of the Moon in great detail, and it is found that the far side is just as rugged, just as cratered and just as lifeless as the hemisphere we have always known.

It is instinctive to assume that full moon is the best time to start observing. Actually, it is the worst! The sunlight is falling 'straight down' onto the lunar surface, and there are practically no shadows; all that can be made out is a medley of light and dark, dominated by the system of bright streaks or rays which issue from some of the craters, notably Tycho in the southern uplands and Copernicus in the Oceanus Procellarum (Ocean of Storms).

The best views are obtained when the Moon is a crescent, half, or gibbous (between half and full). The boundary between the day and night hemispheres is known as the *terminator* (no connection with the film!).

When a crater lies on or near the terminator, its floor will be wholly or partially in shadow, and the effect will be spectacular because the long shadows throw the lunar geography into sharp relief. Later, when the Sun has risen high over the area, that same crater may be hard to identify at all unless it has a very dark floor (as with Plato and Grimaldi) or exceptionally bright walls, as with Aristarchus in the Oceanus Procellarum, which is so reflective that it has often been mistaken for an erupting volcano. The apparent changes in the surface features even over periods of an hour or two are very marked indeed. These are of course caused by the lengthening and shortening of shadows.

Fig. 4.7 The craters on the *left* of the picture, near the terminator, are far easier to see than those on the *right* because of the angle at which sunlight reaches the surface of the Moon

The picture in Figure 4.7 shows how the best view is near the terminator. The craters on the left of the frame are of course on average no deeper than those on the right – it is just that the angle at which sunlight falls on them makes them much easier to see.

Observing the Moon

A very good scheme is to take an outline map, and make a series of drawings of the craters under different angles of illumination. It takes time, but if you persevere you will soon know how to find your way about the Moon. Don't try to draw too large an area at once. For instance, if you are sketching Aristarchus, restrict your drawing to Aristarchus itself and its immediate neighborhood, taking in the crater Herodotus and the huge, winding valley which is one of the most fascinating features on the entire surface.

If you are setting out to make a drawing, begin by using a low magnification, and sketch the main details. Then change to a higher power, and add the more delicate features. Always note the date, time (GMT), telescope used, magnification, and seeing conditions. These are usually given on a scale, due originally to the Greek astronomer E.M. Antoniadi, from 1 (perfect; very rare) down to 5 (very poor).

Features of special interest include the great chains of craters such as the Ptolemaeus group, near the centre of the disk; the Straight Wall, in the Mare Nubium (Sea of Clouds) which is not a wall, but a fault, appearing dark before full moon, because of its shadow, and bright after full, because the sunlight strikes its inclined face; the Sinus Iridum (Bay of Rainbows) which leads off the Mare Imbrium and at sunrise may appear as a 'jeweled handle' as the solar rays strike its mountain border; and the long rilles of Hyginus and Ariadaeus, in the region of the Mare Vaporum (Sea of Vapors).

There is endless detail within the range of your budget telescope, and each one has its own intriguing points. You can also see the 'earthshine,' when the Moon is in crescent stage, and the night side is dimly lit by light sent to it from Earth. To get the best out of observing the Moon, select a small area for a night's observation, and try to see as much detail as possible.

Occasionally the Moon passes into the core of shadow cast by the Earth, and is eclipsed; these lunar eclipses are lovely to watch. When the supply of direct sunlight is cut off, the Moon turns dim, often a coppery color until it emerges. Usually it does not vanish, because some sunlight is bent or refracted onto it by way of the shell of atmosphere surrounding the Earth.

Very mild events can be seen on the Moon, due to trapped gases leaking out from below the crust. These are few and far between, and it has to be said that these Transient Lunar Phenomena (TLP) are too elusive to be detected with small telescopes.

It isn't difficult to take impressive photographs of the Moon with a budget telescope, because the Moon is so bright in the night sky. See Chapters 6 and 7.

The Sun

There's just one golden rule about looking directly at the Sun through any telescope, pair of binoculars or even an SLR camera: ***don't***. Not ever. The Sun's surface is hot, at a temperature approaching 6,000°C. Focus this onto your eye for even a fraction of a second, and you will be blind in that eye for the rest of your life.

Unfortunately this has happened in the past, and a moment's lack of thought can have tragic results.

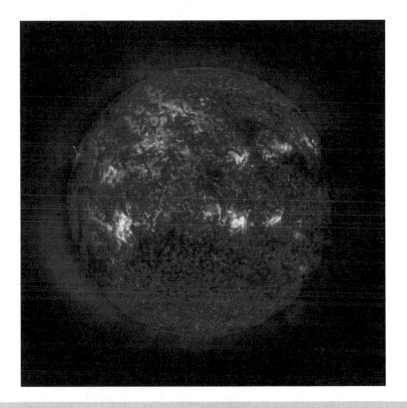

Fig. 4.8 The Sun in ultraviolet: a gigantic exploding hydrogen bomb hanging over our heads!

This image of the Sun was taken by SOHO's EIT (Extreme-Ultraviolet Imaging Telescope) and is reprinted here by courtesy of the EIT Consortium. It was taken on 28th February 2000. It shows rather clearly that the Sun is a boiling mass of energy, vastly violent and constantly changing. When you look at the Sun you are looking straight into a vast nuclear explosion.

So just how do we use our small telescope to observe the Sun safely?

Safety and Finderscopes

Always cap or cover the front of your telescope's finderscope before starting any observations of the Sun. Looking through *any* telescope at the Sun can seriously damage your eyesight, and a finderscope is no exception. Cover the finderscope, because it's all too easy to forget what you're observing and unthinkingly try to use it to aim the main telescope.

You can aim your telescope at the Sun by looking at its shadow. Just point it approximately at the Sun and then make final adjustments so that its shadow on the ground is as small as you can get it.

Solar Filters

Dark filters (such as those that used to be supplied with cheap telescopes) are of no use; they are horribly dangerous. They can never give full protection, and they are always liable to shatter without warning. *If you come across one, throw it away.*

Luckily there are two safe alternatives, one more expensive than the other. The usual, and cheaper, option is to use a *full-aperture solar filter* which fits over the front of the telescope. Such filters are made specifically for the purpose. 'Cheap' is a relative term: for one of a suitable size for our small telescope you will be looking at about £50 ($80), although it may cost somewhat less in the US than in the UK. Solar filters are made of optical glass, with a nickel–chromium, stainless steel, or some similar metallic coating that reflects 99.999% of the sunlight *before* it gets into the telescope tube.

Used properly and with proper precautions full-aperture solar filters are completely safe, but you must make sure that the filter comes from a 'known' source, is in good condition, and is very securely fixed in place. *Never, ever use anything other than a purpose-made full-aperture solar filter for this purpose. Other materials may look similar, but could blind you!*

At the time of writing, in the US and some other countries you can buy less costly filters made of metal coated mylar, a thin, tough plastic. In the UK these are not sold on safety grounds because it is possible for the plastic to be punctured or torn easily, rendering it ineffective – and dangerous – as a solar filter. If you are using a mylar filter, check it for defects *every time you use it*.

Figure 4.10 shows a typical view of the Sun as seen through a 3½-inch (90 mm) telescope equipped with a full-aperture solar filter:

The first details to be seen are the sunspots – always provided that any are present at the time of observation. A sunspot appears as a darker patch on the bright surface or photosphere. It looks black, but this is because of contrast; if it could be seen shining on its own, the surface brightness would be greater than that of an arc-lamp, but only because its temperature is around 2,000° cooler than that of the photosphere. There is usually a black central region or *umbra*, surrounded by a

Fig. 4.9 A full-aperture solar filter – the *only* safe way to observe the Sun with an astronomical telescope

Fig. 4.10 The Sun in white light, showing sunspots on the surface

lighter *penumbra*. Many umbrae can be contained in one *penumbral mass*, and the shapes can be intricate; single spots can appear, but most are members of groups. A typical group has two main members, a leading spot and a follower, with others scattered around. The Sun's surface is gaseous, and so spots are not permanent; they change in shape and complexity from day to day, often from hour to hour, and few last for more than a few weeks, while small spots devoid of penumbra are very short-lived indeed.

Some spot-groups are immense. One, seen in April 1947, covered over 18,000 million square kilometers (over 7,000 million square miles), and was visible with the naked eye – but of course you must never stare straight at the Sun, even when it is low in the sky and looks deceptively harmless.

If you want to look at the Sun with the naked eye (this does NOT apply to using any form of optical aid) you can get special 'Solar Viewing' spectacles, sometimes called 'Eclipse Viewers.' These are inexpensive, but you should buy these either at a specialist astronomical shop or online store, or via one of the (generally American) mail order companies advertising in '*Sky & Telescope*' magazine.

Other 'protective' measures that have from time to time been recommended, such as smoked glass, exposed film, or welder's goggles should NOT be used as they are either dangerous or cannot be relied upon. And remember, solar viewing spectacles are safe for *naked eye* use only.

Sunspots are essentially magnetic phenomena, and form where the Sun's lines of magnetic force below the visible surface break through the photosphere, cooling it. They are not always present. The Sun is to some extent a variable star; it is at its most active every 11 years (approximately), with many spot-groups and associated phenomena, after which activity dies down to a minimum, when there may be many days with no spots at all. When the minimum is past, activity builds up again to the next maximum. The most recent maximum fell in 2000–2001, so 2007 was the 'solar minimum.' Everyone expected the Sun to become more active quite quickly after this, but for some reason the sunspot minimum period lasted years longer than usual. Only now (at the time of writing, 2011) is the Sun becoming visibly active again, with at least some sunspots to be seen on most days.

The Sun spins on its axis, but not quite in the way that a rigid body will do (remember, it is gaseous). The solar equator rotates faster than elsewhere. The equatorial rotation period is approximately 25 days, increasing to rather less than 30 days at mid-latitudes and as much as 34 days at the poles – though in fact spots are never found either in the polar zones or very close to the equator.

Watch the Sun from day to day, and you will see how the spots are carried across the disk by virtue of the solar rotation. It will take a group around a fortnight to cross from one limb to another, and subsequently it will reappear at the opposite limb, provided that it still exists. Some groups persist for several crossings, the longevity record so far is held by a group which lasted for 200 days, between June and December 1943. On the other hand, a very small spot (a 'pore') maybe gone in an hour. When near the limb, a sunspot is foreshortened, because the Sun is a globe; with a regular spot it is seen that the penumbra toward the limb is broadened with respect to the penumbra on the opposite side. This was first moved by A. Wilson as long ago as 1774, and from it he inferred that a spot must be a depression rather than a bulge. The 'Wilson effect' can be very obvious, though not all spots show it.

Our budget telescope, fitted with a solar filter, will also show faculae (Latin, 'torches') which are bright, active regions above the surface composed mainly of hydrogen. They are associated with major spot-groups, but may also be seen on their own, generally in areas where a spot-group has disappeared or where a spot is about to break out.

Solar Eclipses

A solar eclipse occurs when the new moon passes in front of the Sun, blocking out the photosphere. A total solar eclipse is a magnificent sight; the sky darkens, the planets and bright stars come out, and the black disk of the Moon is surrounded by the solar atmosphere, known as the corona. There may also be prominences, masses of red, glowing hydrogen rising from the Sun's surface.

A solar eclipse can be watched with the naked eye, protected by the 'Solar Viewing' spectacles ('Eclipse Viewers') mentioned above. If you want to have an Eclipse Party (a great excuse for a barbecue!) buy several pairs of viewers so that your friends can join in: they cost only a couple of pounds (or dollars) a pair.

In England, the last total eclipse occurred in 1999 and the next will not be until 2090, though other parts of the world are more favored. No telescope is needed to enjoy the beauty of a total eclipse; much the best instrument is the naked eye. If the Sun is not completely hidden, the corona and prominences cannot be seen, but observations can be made with a telescope – fitted, of course, with a solar filter.

All in all, the Sun is one of the very best targets for our budget telescope, but always take care: a cat may look at a king, but no telescope-user should ever look directly at the Sun!

Specialist Solar Telescopes

These are not strictly 'budget' instruments, but at least these amazing solar telescopes are getting less expensive! They are worth a mention here. The Coronado PST (Personal Solar Telescope) shown below is the least expensive. Solar telescopes are designed for solar observing at a specific, very narrow, wavelength called the hydrogen-alpha line. This is a specific red spectral line at wavelength of 656.28 nm, which is the frequency of the light emitted when a hydrogen electron falls from its third to its second lowest energy level. PSTs currently costs about $600 in the USA, but for some reason about £500 in the UK, considerably more. (No, we don't know why.) They are available used for about half this amount. Two things to be careful about when buying a used PST. The first is obvious: safety. If there is any damage to the telescope or evidence that it has been disassembled at any time, don't buy it. The second point is that early PSTs had a gold colored anti-reflection coating on the objective lens, and this coating was prone to serious deterioration which affected the image contrast. Later models have a bluish coating (it looks like the coating on a camera lens), and are fine.

Fig. 4.11 Coronado PST – at the time of writing, the least expensive hydrogen-alpha solar telescope

A hydrogen-alpha solar telescope enables an observer to see solar prominences, filaments, active regions of the Sun, and some surface detail as well. Figures 4.12 and 4.13 show typical views of some solar prominences, as seen (and photographed) through a PST.

Fig. 4.12 A typical view of the Sun, observed with the PST

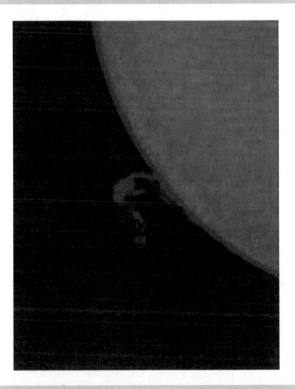

Fig. 4.13 A gigantic solar flare, observed with the PST. This view lasted only a couple of hours before the flare dissipated
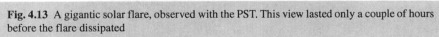

The Sun's surface details as seen from Earth change very rapidly, and during the course of an afternoon's observing it is possible to see prominences develop, grow, and collapse again.

The Planets

As we said before, don't expect too much! Of all the planets, our budget telescope will give good results only on Venus, Mars, Jupiter and Saturn. Of the rest, Mercury is always inconveniently close to the Sun in the sky, and Uranus and Neptune are so remote that they can easily be mistaken for faint stars, while Pluto is too dim to be seen at all. This isn't a failing only of small telescopes – even the largest amateur telescopes cannot show detail on any of the planets beyond Saturn, nor anything more useful than the phases of Mercury.

That said, it is still fascinating to see surface details on the planets for yourself.

Mercury

Mercury is the nearest planet to the Sun, ferociously hot, and rather Moon-like in its surface appearance.

The mosaic image of Mercury shown in Figure 4.14 below was taken by the Mariner 10 spacecraft during its approach on 29 March 1974. It consists of 18 images taken at 42-second intervals during a 13 minute period when the spacecraft was 125,000 miles (200,000 km) from the planet.

With the naked eye, Mercury can be glimpsed only when at its best, either very low in the west after sunset or very low in the east before dawn. *Using a budget telescope – or any other telescope that isn't computer-controlled – to sweep for it when the Sun is above the horizon is emphatically not to be recommended. Failure is almost certain, and there is always the danger of looking at the Sun by mistake.*

Mercury shows lunar-type phases from new to full, but even these are not easy to see with a small telescope, and any surface detail is completely out of range. All in all, the only real pleasure of observing Mercury is success in finding it!

Fig. 4.14 Mercury: a mosaic image taken by Mariner 10

Venus

Venus is a hot, cloudy planet with a very dense atmosphere. Without going there, we can see only the tops of the clouds. Its atmosphere is made up chiefly of carbon dioxide; the clouds are rich in sulfuric acid, so that Venus is a most unfriendly place. Here, in Figure 4.15, is an image of Venus, taken in ultraviolet light by the Pioneer Venus Orbiter on February 5, 1979.

The swirling weather patterns can be seen clearly, but little else. The surface of Venus, at least as much of it as has been seen by our space probes, looks like a boulder-strewn desert.

Seen from Earth, Venus shows phases like the Moon and Mercury, because it is nearer to the Sun than we are. The different phases are fairly easy to make out with a small telescope and it is well worth looking for them.

But although Venus is much the brightest object in the sky apart from the Sun and the Moon, it is telescopically disappointing. The surface is permanently hidden by its thick, cloud-laden atmosphere.

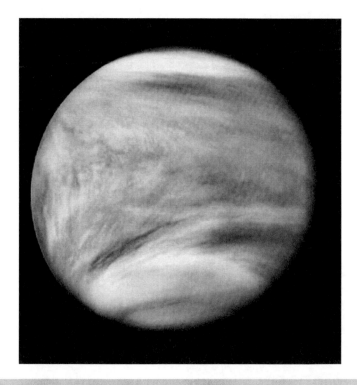

Fig. 4.15 Weather patterns on Venus

Finding Venus is easy. It is always near the Sun in the sky, and is almost always the brightest object, by quite a large margin.

In a small telescope, Venus looks like Figure 4.16.

Fig. 4.16 Venus, as seen with a budget telescope

This is the 'crescent Venus', and is the point at which it is easiest to see that Venus exhibits phases. The main interest in observing Venus is in following the changing phase, which is easy enough, though everything happens at a very leisurely pace. The 'synodic period' – that is to say, the interval between successive inferior conjunctions – is nearly 584 days. An 'inferior conjunction' is when it is between the Earth and the Sun, so that its night side is towards us.

If the alignment is perfect, Venus will appear in *transit* as a dark disk crossing the Sun. Don't hold your breath – it doesn't happen very often. A transit of Venus last happened in 1882, then in June 2004, and then again in June 2012. After that, there won't be any more until 2017 and 2125, by which time you probably won't care.

The moment of *dichotomy* (half-phase) of Venus can be worked out very accurately, but it always happens during evening apparitions. When Venus is waning, dichotomy is early by a day or two; when Venus is waxing in the morning sky, dichotomy is late. This was first noticed in the 1790s by the German astronomer Johann Schröter, and years ago I (Patrick) christened it the Schröter effect, a term which has now become widely used. It is due to the effects of Venus' dense, extensive atmosphere.

It sometimes happens that during the crescent stage, the night hemisphere of Venus shines dimly. Venus has no satellite, and so this '*Ashen Light*' cannot be analogous to Earthshine on our Moon; it seems to be due to electrical effects in the planet's upper atmosphere. However, it is always elusive. There is no record of its being seen with a small amateur telescope, but is always worth looking for. You could be the first!

Surface markings on Venus are always very vague, and are due to cloud phenomena. In most cases the disk will appear completely blank.

Mars

Mars is much more Earth-like than either Venus or Mercury. It is colder than Earth, being further from the Sun. It is intermediate in size between the Earth and the Moon. It has a thin atmosphere.

Fig. 4.17 Mars, as imaged by the Hubble Space Telescope at the June 2001 opposition

It also has seasons.

This global image was taken by the Wide-Field Planetary Camera 2 on the Hubble Space Telescope. It shows a region of Mars known as the Syrtis Major region. At the top of the picture, the polar ice cap – which changes size markedly with the seasons – is clearly visible.

The surface of Mars has been imaged from close up. See Figure 4.18.

Fig. 4.18 The surface of Mars, taken from the surface of Mars!

When a planet is exactly opposite to the Sun in the sky, it is said to be at opposition; it is due south at midnight (as seen from the Earth's Northern Hemisphere) and is best placed for observation. Mars comes to opposition every other year; oppositions occurred in April 1999, June 2001, and August 2003, but none in 2000 or 2002. In 2003 Mars was at its closest, a mere 35,000,000 miles (56,000,000 km) from the Earth, and will appear as a brilliant red object, outshining all the planets apart from Venus. There will be oppositions in 2014, 2016, 2018 and 2020.

Mars is not easy to study with a small telescope. It's best to wait until it is fairly near opposition – say 5–6 weeks to either side of the opposition date, but of course it is worth a look at other times. Fortunately the Martian atmosphere is so thin that it does not usually mask the surface features, though major dust-storms can occur, as they did in September/October 2001.

Under good conditions, our budget telescope will show the polar ice-caps and the main dark areas (once thought to be seas but now known to be regions where winds in the tenuous atmosphere have blown away the red, dusty material which covers most of the planet, exposing the darker layers below). The surface markings, unlike those of Venus are permanent (because all we can see on Venus is the tops of the clouds), and have been named. The two most prominent are the Syrtis Major in the equatorial region of the planet and Acidalia Planitia in the north; the Syrtis Major is V-shaped, and much the easiest to make out.

Figure 4.19 shows the small telescope view of Mars, such as you might be able to see on a perfect night with one of the best budget-price instruments:

Fig. 4.19 Mars as it should appear in a budget telescope in the best seeing conditions

Begin by looking for surface detail – any surface detail at all – and then go on to see if you can detect changes in size of the ice cap over time.

Hellas, an impact basin in the Southern Hemisphere, is sometimes cloud-filled, and can then look very like a second polar ice-cap in small telescope.

Don't expect to be able to make out a lot on Mars the first time you look though your telescope. Observing takes lots of practice, and you will find that many hours of experience will make an enormous difference to what you can make out.

Train your telescope on Mars, make out what is on view and then do your best to draw it. Occasionally, when the sky is perfectly transparent and Mars is well placed, you will be astounded by how much is visible, but don't let a succession of nights of average seeing and the repeated sight of Mars as a slightly fuzzy ball put you off completely. Persevere!

Mars has two moons, but both the tiny Martian satellites, Phobos and Deimos, are hopelessly out of the range of most amateur telescopes, and certainly budget ones.

Jupiter

Jupiter is a giant world, nearly 89,000 miles (140,000 km) in diameter as measured through its equator (it isn't quite spherical; it is flattened at the poles), and moves round the Sun at a mean distance of 483,000,000 miles (775,000,000 km), in a period of almost 12 years.

Fig. 4.20 Jupiter, the solar system's largest planet, imaged by the Hubble Space Telescope

Jupiter's surface is made up of gas, mainly hydrogen with a good deal of helium; below there are layers of liquid hydrogen, overlying what is probably a silicate core. The inner temperature may exceed 30,000°, but this does not mean that Jupiter is a miniature sun – and the upper clouds are bitterly cold.

Jupiter is affected by huge storms and cyclones that come and go over periods of months or years. The Great Red Spot, near the bottom of the image in Figure 4.21, is the largest known storm in the Solar System. With a diameter of 15,400 miles, it is almost twice the size of the entire Earth and one sixth the diameter of Jupiter itself.

Fig. 4.21 The Great Red Spot – a gigantic storm system

The Red Spot does change its shape, size, and color, sometimes dramatically. The picture was obtained by NASA's Hubble Space Telescope.

Once you have located Jupiter, in the southern sky (from the northern hemisphere), it is easy to find again because of its distinctive appearance. To the naked eye it is quite red, and very obviously different from any other celestial objects.

Seen through a budget telescope, the giant of the Sun's family is much more rewarding than any of the inner planets. It comes to opposition (its closest to Earth) every 13 months, and is therefore well placed for observation for several months in every year. It is outshone only by Venus and, very occasionally, by Mars.

Jupiter's yellow disk is obviously flattened, because the axial rotation period is very short (less than 10 hours) and the equator bulges out. Obviously the surface details shift and change, but there are always several 'cloud belts', of which the two most prominent lie to either side of the Jovian equator. Our small telescope will show several other belts under good seeing conditions. There are also other features, including spots. The Great Red Spot has been under observation on and off (more on than off) for several centuries!

Begin observing Jupiter with a low power, and change to higher magnification to see the finer details. The rapid rotation means that surface features such as the spots are carried across the disk with surprising speed; the shifts can be noticed after only a few minutes, so if you intend to make a sketch, do not take too long about it. And, as usual, remember to note the observational details. Times should be given to the nearest minute.

Jupiter has a wealth of satellites, but only four are within the range of small telescopes; these are Io, Europa, Ganymede and Callisto, known collectively as the Galilean satellites, or just 'Galileans' because they were first closely studied by the great Italian scientist Galileo in 1610.

Here is a budget telescope view of Jupiter, with its four Galilean moons – left to right in the picture in Figure 4.22, these are Callisto, Europa, Ganymede and Io:

Fig. 4.22 Jupiter and four of its moons, as seen with a budget telescope

Europa is slightly smaller than our Moon, Io slightly larger, and Ganymede and Callisto much larger; indeed, the diameter of Ganymede is greater than that of the planet Mercury.

All four have been surveyed from close range by spacecraft; Ganymede and Callisto are icy and cratered, Europa icy and smooth, and Io violently volcanic, with eruptions going on all the time. With our telescope however, they will look like stars. But you can easily see them moving, in the course of a single night. They move round Jupiter in periods ranging from less than 2 days (Io) up to 16½ days (Callisto).

The Galileans produce all sorts of phenomena. They may pass in transit across Jupiter's disk; there are also shadow transits, and with a large (amateur) telescope the black spots of shadow are clearly visible. The satellites may pass behind Jupiter, and be hidden or occulted; they may pass into Jupiter's shadow, and be eclipsed. It is fascinating to follow these antics, and a low-cost telescope is fully adequate for the task. Jupiter and its system provide a constant source of interest.

Saturn

Beyond Jupiter, at a mean distance from the Sun of 886,000,000 miles (1,427,000,000 km) comes the ringed planet Saturn, arguably the most beautiful object in the whole of the sky.

Fig. 4.23 Saturn, imaged by the Hubble Space Telescope

NASA's Voyager 2 took this photograph of Saturn on July 21, 1981, when the spacecraft was 21,000,000 miles (33,900,000 km) from the planet. The moons Rhea and Dione appear as dots to the south and southeast of Saturn, respectively.

Saturn, like Jupiter, is well seen in a telescope for several months in every year; it has a long orbital period (29½ years) but a short 'day' of only 10¼ hours, so that its disk is strongly flattened. There are cloud belts, much less obvious than those of Jupiter, though our telescope should show one or two of them on a good night. Spots are rare and temporary; the most prominent white spot of the last century was that of 1933, discovered by the amateur astronomer W. N. Hay (perhaps better remembered as Will Hay, the stage and screen comedian). It was clearly visible with a 3-inch telescope, and persisted for some weeks.

Look for Saturn in the southern sky (in the northern hemisphere). As with Jupiter, you will remember its appearance once you have found it for the first time. It is distinctly yellow (not red like Jupiter) and usually quite bright.

The glory of Saturn lies in its ring system, which you can see clearly with a small telescope, see Figure 4.24.

Fig. 4.24 Saturn, as seen with a budget telescope

There are several main rings, made up of swarms of thousands of millions of icy particles moving round the planet in the manner of dwarf moons. Our telescope will show two main rings, A and B, and under good conditions the gap which separates them, known as Cassini's Division in honor of the Italian astronomer who discovered it as long ago as 1675. The rings are extensive, measuring around 170,000 miles (270,000 km) from tip to tip but they are also very thin, and when placed edgewise-on to us, as was the case in 2009, they disappear from view in small telescopes.

However, our view of the rings of Saturn is getting more favorable. By 2016 the rings will be at their most 'open', and will be spectacular. As they open (that is, our view of them becomes more from 'above' and less 'edge-on') the Cassini Division will be obvious, and also the shadow of the rings on the globe – and the shadow of the globe on the rings.

Saturn is an awkward object to sketch; to make a really good drawing you need at least reasonable artistic ability.

Saturn's satellite system differs from that of Jupiter. There is one large moon, Titan, which is bigger than Mercury, and it is the only satellite in the Solar System known to have a dense atmosphere; any telescope will show it. A 3-inch (75 mm) or larger telescope should also be able to pick up Rhea, Iapetus, Dione and Tethys, but not easily. All the remaining satellites are much fainter.

Saturn is smaller than Jupiter (equatorial diameter 75,000 miles, 12,000 km) and is also much further away. It is therefore less bright than Jupiter, even though it can outshine all the stars apart from Sirius and the far-southern Canopus.

Uranus and Neptune

Observing these objects is not for beginners. With almost any amateur telescope they will appear as faint star-like dots.

These outer giants are around 30,000 miles (48,000 km) across. Uranus is just visible with the naked eye if you know where to look for it, but to see Neptune you require optical aid, either binoculars or a telescope.

Uranus is fairly easily found if you have an adequate star-map; our typical small telescope will show that it looks rather larger and dimmer than a star. Neptune can also be located without much trouble, but in our budget-priced telescope it will look just like a faint star. You will need a reasonably detailed star-map or planetarium program if you want to identify it.

Both these outer planets are slow movers; Uranus takes 84 years to orbit round the Sun, while Neptune requires nearly 165 years. Neptune was discovered only in 1846; this is less than one Neptunian year ago!

Pluto, as we have said, is out of range for low-cost telescopes. To see it at all, even as a tiny point, you need an aperture of at least 8 inches (200 mm).

The Asteroids

The asteroids are very small bodies, most of which keep to the gap between the orbits of Mars and Jupiter, though very minute members of the swarm may swing inward and approach the Earth. Only one asteroid (Vesta) is ever visible with the naked eye, but there are quite a number within a range of budget-priced telescopes.

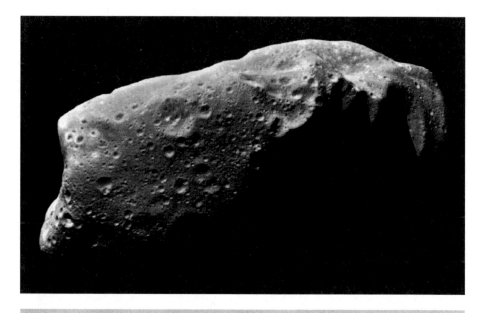

Fig. 4.25 Asteroid 243 Ida, as imaged by the Galileo spacecraft from a couple of thousand miles away

This view of the asteroid 243 Ida is a mosaic of five image frames acquired by the Galileo spacecraft's solid-state imaging system at ranges of 1,900–2,375 miles (3,057–3,821 km) on August 28, 1993.

Ida was the second asteroid ever encountered by a spacecraft. It appears to be about 32 miles (52 km) long, and is clearly irregularly shaped. It is believed to be like a stony or stony-iron meteorite in composition.

Positions of the asteroids are given in yearly almanacs, and they can be tracked down fairly easily if a good star-map is available, but of course they look exactly like stars.

We wouldn't recommend asteroids as targets for beginners, either.

Comets

Comets are the most erratic members of the Sun's family. A typical large comet has a *nucleus*, made up of ices mixed in with rocky particles; a gaseous head or *coma*, which develops only when the comet is comparatively close to the Sun; and a *tail* (or tails), made up of tiny particles and tenuous gas. Really bright comets are rare; the last was Hale-Bopp, which remained on view for months in 1997.

Most comets have very eccentric orbits; for example Hale-Bopp will not return to the inner Solar System for over 2,000 years. There are however many short-period comets, which return to perihelion regularly; Encke's Comet has a period of only 3.3 years. All these are faint, and few develop tails; telescopically they look rather like very dim patches of cotton–wool in the sky. Their movements against the starry background can be noticed even after only an hour or two.

The photograph in Figure 4.26, showing the nucleus of comet P/Halley was taken with the Halley multicolor Camera on the ESA mission spacecraft Giotto, which passed within 569 km of the comet on July 10, 1992.

Fig. 4.26 Comet P/Halley, imaged by the Giotto space probe, less than 500 miles from the comet

Although amateur astronomers have been very successful in discovering new comets, an inexpensive telescope is not suited to this sort of work; you need a wide field of view, and good light-grasp. You also need hours and hours of spare time! The late George Alcock, one of the most successful comet-hunters of modern times, used a pair of large-aperture, specially mounted binoculars; so far as we know, he never owned a telescope in his life!

Figure 4.27 shows a (non-telescopic) view of Comet P/Halley and was taken on March 8, 1986 by W. Liller, Easter Island, as part of the International Halley Watch (IHW) Large Scale Phenomena Network.

Fig. 4.27 An earth-based photograph of comet P/Halley, taken in 1986

Chapter 5

Observing the Stars and Galaxies

Come now to the stellar heavens, where there is a great deal of scope for the user of a budget-priced telescope.

As mentioned earlier, the first step is to learn your way round the sky, which is not nearly so difficult as might be thought, because the constellation patterns do not change, and the planets are easily found (only Saturn and, at its faintest, Mars can ever be mistaken for stars).

Stars are graded into 'magnitudes' of apparent brilliancy. The lower the magnitude, the brighter the star. Normal-sighted people can see down to magnitude 6 without optical aid; modern electronic devices used with large telescopes can reach down to magnitude +30. At the other end of the scale, four stars have minus magnitudes; Venus reaches −4, the full moon −14, and the Sun −27.

Note that a star's apparent magnitude is not a reliable key to its real luminosity. A star may look bright because it is genuinely very powerful, because it is relatively close, or a combination of both. The two brightest stars in the sky are Sirius, in the constellation of Canis Major (the Great Dog) and the far-southern Canopus, in Carina (the Keel). Their magnitudes are respectively −1.4 and −0.7, so that Sirius is considerably the brighter of the two. Yet Sirius at a distance of 8.6 light-years, is only 26 times as powerful as the Sun; according to the authoritative Cambridge Catalog Canopus, at 1,200 light-years, could match 200,000 Suns. (One light-year is the distance travelled by a ray of light in 1 year; rather less than six million million miles (and over nine million million kilometres).

Stars are given *catalog numbers*; according to the system introduced in 1603 by the German astronomer Bayer, stars in each constellation are given Greek letters, theoretically in order of brilliancy, starting with Alpha and working through to Omega – though in practice the order is frequently chaotic, and of course the system is limited to 24 stars (the number of letters in the Greek alphabet); fainter stars are

P. Moore and J. Watson, *Astronomy with a Budget Telescope: An Introduction to Practical Observing*, Patrick Moore's Practical Astronomy Series, DOI 10.1007/978-1-4614-2161-0_5, © Springer Science+Business Media, LLC 2012

given numbers. Many stars have individual names, usually Arabic but in general these names are used only for stars of the first magnitude, plus a few special cases.

Here is the Greek alphabet:

Letter	Lower case	Upper case
Alpha	α	A
Beta	β	B
Gamma	γ	Γ
Delta	δ	Δ
Epsilon	ε	E
Zeta	ζ	Z
Eta	η	H
Theta	θ	Θ
Iota	ι	I
Kappa	κ	K
Lambda	λ	Λ
Mu	μ	M
Nu	ν	N
Xi	ξ	Ξ
Omicron	o	O
Pi	π	Π
Rho	ρ	P
Sigma	ς	Σ
Tau	τ	T
Upsilon	υ	Y
Phi	φ	Φ
Chi	χ	X
Psi	ψ	Ψ
Omega	ω	Ω

Locating Stars

The 'finder charts' in this book should help you, but only once you are familiar with the night sky to some extent. There is no substitute for experience, but the best starting point is either a good star atlas, planetarium software (there is freeware available on the Internet), or a planisphere (a simple card or plastic device that shows the heavens for any particular time and date; unfortunately planispheres work at only a given latitude). Monthly star charts showing the constellations are published in the astronomical magazines, and even in some newspapers.

What you need to do is familiarize yourself with the constellations. Once you can find the principal stars in the main constellations, you will find locating individual objects is easier than you first thought!

The most useful technique for finding faint objects is 'star hopping.' Start by aiming the telescope at a 'known' or easily found object, and simply 'hop' from star to star until you find what you are looking for. Like everything, it takes practice but is very effective once you have got the hang of it.

It may be useful to give a few 'typical stars' for each magnitude:

Magnitude	Stars	Constellation
0	Vega (Alpha Lyrae), 0.0	Lyra (the Lyre or Harp)
	Capella (Alpha Aurigae), 0.1	Auriga (the Charioteer)
	Rigel (Beta Orionis), 0.1	Orion
1	Aldebaran (Alpha Tauri), 0.9	Tauris (the Bull)
	Antares (Alpha Scorpii), 1.0	Scorpius (the Scorpion)
	Spica (Alpha Virginis), 1.0	Virgo (the Virgin)
1½	Regulus (Alpha Leonis), 1.4	Leo (the Lion)
	Adhara (Epsilon Canis Majoris), 1.5	Canis Major (the Great Dog)
	Castor (Alpha Gemiorum), 1.6	Gemini (the Twins)
2	Alphard (Alpha Hydrae), 2.0	Hydra (the Watersnake)
	Polaris (Alpha Ursae Minoris), 2.0	Ursa Minor (the Little Bear)
	Hamal (Alpha Arietis), 2.0	Aries (The Ram)
2½	Phad (Gamma Ursae Majoris), 2.4	Ursa Major (the Great Bear)
	Alderamin (Alpha Cephei), 2.4	Cepheus
	Markab (Alpha Pegasi), 2.5	Pegasus
3	Sadalmelik (Alpha Aquarii), 3.0	Aquarius (the Water Carrier)
	Albireo (Beta Cygni), 3.1	Cygnus (the Swan)
	Tais (Delta Draconis), 3.1	Draco (the Dragon)
3½	Altarf (Beta Cancri), 3.5	Cancer (the Crab)
	Rana (Delta Eridani), 3.5	Eridanus (the River)
	Adhafera (Zeta Leonis), 3.4	Leo (the Lion)
4	Asellus Australis (Delta Cancri), 3.9	Cancer (the Crab)
	Alkhiba (Alpha Corvi), 4.0	Corvus (the Crow)
	Nembus (Upsilon Persei), 4.0	Perseus
4½	Thabit (Upsilon Orionis), 4.5	Orion
	Yildun (Delta Ursae Minoris), 4.5	Ursa Minor (the Little Bear)
	Kappa Aurigae, 4.4	Auriga (the Charioteer)
5	Chi Cancri, 5.1	Cancer (the Crab)
	Zeta Sagittae, 5.0	Sagitta (the Arrow)
	Eta Ursae Minoris, 4.9	Ursa Minor (the Little Bear)

All these stars can be identified by means of a reasonably detailed map or planetarium software. Look on the Internet where you can find 'freeware' planetarium software, or check out www.Springer.com/sky and www.Amazon.com for suitable maps and books.

Star Colors

Many people assume that all stars are of the same color. In fact, nothing could be further from the truth. The Sun is a yellow star; Sirius and Altair are white; Betelgeux and Antares are orange-red, and so on. These differences in color are due to real differences in surface temperature. The Sun's surface is at a temperature of between 5,000°C and 6,000°C, while that of Sirius is about 11,000° and that of Betelgeux a mere 3,000°. To make up for this, Betelgeux is huge; its globe could contain the whole orbit of the Earth round the Sun, and it has 15,000 times the Sun's luminosity.

These colors can be seen faintly with the naked eye, but are much better brought out with a telescope.

Here we have to agree that reflectors are more reliable than refractors. Starlight is made up of all the colors of the rainbow, and when a ray of light is passed through a lens the different parts of it tend to split up, producing undesirable false color. A poor-quality lens will yield a star image surrounded by gaudy rings which may look attractive, but are most unwelcome to the observer! Even a good refractor cannot be entirely free from false color, though compound object-glasses reduce it to an acceptable level. And in absolute terms, some of the telescopes we are considering may be excellent value, but are unlikely to contain the best quality optics.

On the other hand, a mirror reflects all wavelengths equally, and so the only false color in a reflector must come from the eyepiece or the atmosphere. Don't expect the colors of stars to appear other than pastel.

Spectral Classes

Just as a telescope collects light, so a spectroscope splits it up, and an analysis of the light tells us what substances are present in the light-source. Stars are divided into various spectral types, denoted by letters of the alphabet. The main types are:

Type	Color	Typical stars	Surface temperature (°C)
W	Bluish	Regor (Gamma Velorum)	Up to 80,000 Rare
O	Bluish-white	Suhail Hadar (Zeta Pappis)	35,000–40,000 Rare
B	Bluish-white	Rigel, Regulus, Spica	12,000–26,000
A	White	Sirius, Vega, Regulus	7,500–11,000
F	Yellowish	Procyon, Polaris, Canopus	6,000–7,500
G	Yellow	The Sun, Capella	4,200–5,500
K	Orange	Arcturus, Aldebaran, Pollux	3,000–5,000
M	Orange-red	Betelgeux, Antares	3,000–3,400
R	Red	V Arietis	2,600 Rare
N	Red	R Leporis	2,500 Rare
S	Red	Chi Cygni	2,600 Rare

Visually, the only obvious colors are those of the orange and red stars. Vega, even though of type A, is the bluest of the first-magnitude stars, while Antares is definitely the reddest; its name means 'the Rival of Mars' (Ares). Of all naked-eye stars, the only one said to be green in color is Beta Librae or Zubenelchemale, magnitude 2.6, but most people will call it white!

All the reddest stars, of types R, N and S, are variable, and only one (Chi Cygni) is ever bright enough to be seen with the naked eye, though many are within the range of modest telescopes when near maximum. Types R and N are now often combined into Type C.

The Finder Charts in This Book

The full-page finder charts in this book show the positions of celestial objects. They are all on the same scale. The constellations are shown, but of course the orientation of the chart will not necessarily be the same as the orientation of the sky when you are observing. The shaded areas show the Milky Way, which is simply our own galaxy, looking into the dense areas towards the core. Finding your way around should not be too hard, although there is unfortunately no substitute for learning where the constellations are!

Double Stars

Double stars are very common in the sky. They are of two types: *optical pairs*, and *binaries*.

An optical pair (or optical double) is merely a line-of-sight effect, with one star 'in the background', so to speak. A good example is provided by Al Giedi or Alpha Capricorni, in the constellation of the Sea-Goat. The two components are separated by almost 380 seconds of arc (a second of arc is 1/3,600 of a degree), and can easily be seen individually in almost any small telescope. The brighter component, of magnitude 2.9, is 49 light-years away, but the fainter component, magnitude 4.2, lies at 1,600 light-years. There is absolutely no connection between the two stars, and if observed from a different vantage point in the Galaxy they could well lie on opposite sides of the sky.

However, most pairs are *binary systems*, so that the components move round their common centre of gravity just as the bells of a dumbbell will do when twisted by the bar joining them. The separations and the revolution periods vary widely.

Some pairs are so close that a large telescope is needed to show them individually, and there are indeed many binaries which appear single even in the world's most powerful telescopes. On the other hand, there are pairs so widely separated that their revolution periods amount to many centuries. The stars Alnitak or Zeta Orionis, the southernmost of the three stars of Orion's Belt, is made up of two components, of magnitudes 1.9 and 4.0; the orbital period is 150 years.

Some binaries are perfect 'twins' such as Alya or Theta Serpentis in the Serpent. In other cases the components are widely unequal; the brilliant Rigel, in Orion, has a companion of magnitude 7.

Many doubles are within range of budget telescopes. It is not easy to give definite values for limiting magnitudes and separations, because so much depends upon individual observers, but it is possible to give a general guide:

Aperture of telescope		Limiting	Smallest separation,
inches	cm	magnitude	seconds of arc
2	2.5	10.5	2.5
3	7.6	11.4	1.8
4	10.2	12.0	1.3
5	12.7	12.5	1.0

The third column refers to components which are equal. Where the components are unequal, the double will naturally be a more difficult object, particularly when one component is much brighter than the other (so that the dimmer one may be lost in its glare). The *separation* indicates how far apart the components are, or rather seem to be. It is measured in seconds of arc and is in effect the angle made between imaginary lines drawn from each star to your telescope. Put another way, it is the angle through which you would need to move the telescope from pointing exactly at the first star to point exactly at the second. Remember, it depends on line-of-sight and has nothing at all to do with how far apart the stars actually are.

The *position angle* is a measurement you will sometimes come across. It is given by the angle between the primary component and the secondary from 0° (North), through East, South and West, as shown in the diagram (which illustrates a position angle of about 240°).

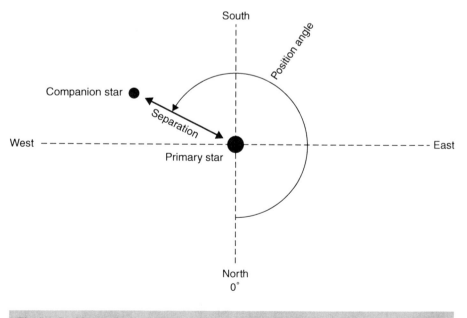

Fig. 5.1 Position angle

Notice the astronomical convention of having South at the top and North at the bottom. This convention is not used in astronautics, but is helpful for astronomy because Newtonian and refracting telescopes invert the image.

List of double stars are given in many books, so we thought the best course here would be to give a few selected examples.

(a) Separable with a 2-inch (50 mm) telescope

Beta Cygni (Albireo). Magnitudes 3.1 and 5.1, separation 34″.4, PA 054°. The faintest of the five stars making up the X of Cygnus. The primary is golden (type K), while the companion is vivid blue. This is unquestionably the loveliest colored double in the entire sky. The primary is 700 times as luminous as the Sun, and the secondary could easily outmatch the Sun. The distance from us is 390 light-years, so that we see Albireo now as it used to be at about the time that Galileo was making the first telescope observations.

In our small telescope it looks like this, although no photograph can do justice to the brilliant, jewel-like appearance of the two stars, nor exactly convey the way the color difference appears. See Figure 5.2, and Chart 5.1.

Theta Serpentis (Alya). Easy to find, not far from Altair in Aquila. Magnitudes 4.5 and 4.5; separation 22″.4; PA 104°. A pair of perfect twins, each 12 times as luminous as the Sun and pure white (type A). They share common proper motion in space, but the revolution period must be immensely long. See Chart 5.2.

Gamma Leonis (Algieba). In the Sickle of Leo, near Regulus. Magnitudes 2.2 and 3.5, separation 4″.6, PA 125°. Binary, but the period is 619 years, so that there is only a very slow change in PA. The primary star is orange (Type K) and the secondary yellowish (G). See Chart 5.3.

(b) Separable with a 3-inch (75 mm) telescope.

Alpha Ursae Minoris (Polaris). The Pole Star, Magnitudes 2.0 and 9.0, separation 18″.4, PA 2180. With our 3-inch, the problem here is the relative faintness of the secondary. It has been claimed that a 2½-inch telescope will suffice, but it is certainly a severe test for any telescope below 3 inch aperture. See Chart 5.4.

Delta Geminorum (Wasat). In the Twins. Magnitudes 3.5 and 8.2, separation 5″.7, PA 225°. Not too easy. The primary is yellowish, the companion bluish. See Chart 5.5.

Gamma Andromedae (Almaak). The end member of the three stars of Andromeda leading off the Square of Pegasus. Magnitudes 2.3 and 5.0, separation 7″.5, PA 001°. The primary is orange (type K) and the companion white (type A), though it often appears bluish because of the effect of contrast. The secondary is itself a binary, with a period of 61 years, but the separation never exceeds 0″.5 – too close for a small telescope. See Chart 5.6.

Alpha Geminorum (Castor). The fainter of the two Twins. Magnitudes 1.9 and 2.9, separation 3″.9, PA 064°. Both stars are white. Castor is a binary, revolution period 420 years. At present the angular separation is slowly increasing. See Chart 5.7.

Chart 5.1 Beta Cygni

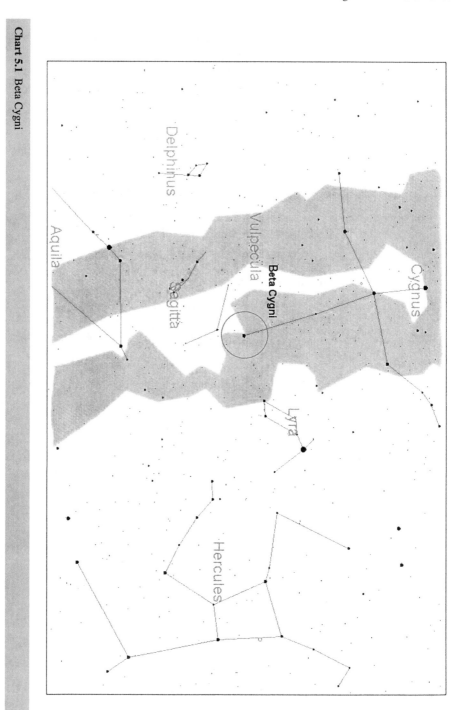

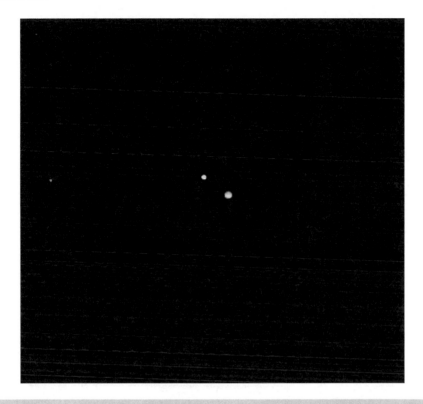

Fig. 5.2 Albirco, showing the dramatic color difference between the two components of this binary star

(c) Separable with a 5-inch (130 mm) telescope:

Gamma Virginis (Porimma – also (but less commonly) known as Arich). The star at the base of the 'bowl' of Virgo. Magnitudes 3.5 and 3.5, present separation 2″, PA 287°. Identical twins, but the pair is not so easy as it used to be! It is a binary with a period of 169 years. A few years ago this binary could be seen only as a single star because the two components were 'in line' in their orbit, but since 2005 they have been moving apart as seen from Earth. See Chart 5.8.

Zeta Aquarii. (Deneb el Okab). Below (south of) the Square of Pegasus. Magnitudes 4.3 and 4.5, separation 1″.9, PA 188°, both stars white. Binary 856 years, the apparent separation is very slowly increasing. See Chart 5.9.

Alpha Piscium (Alrescha). Aries/Pegasus area. Magnitudes 4.2 and 5.1, separation 0″.9, PA 279°. A very slow binary (period 933 years). Fairly easy with a 4-inch; a very difficult test for a 3-inch telescope. See Chart 5.10.

Chart 5.2 Alya

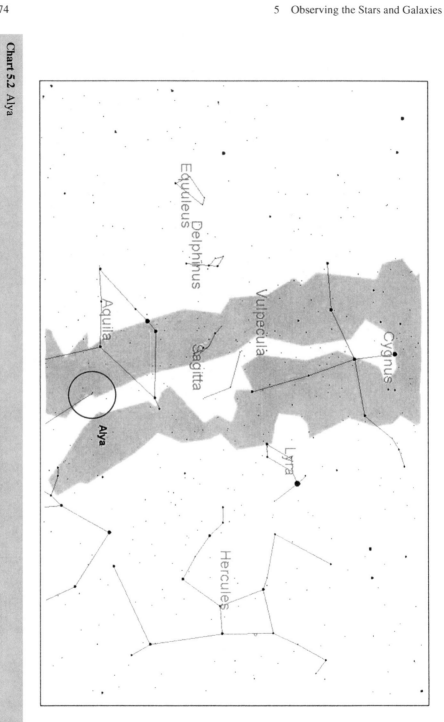

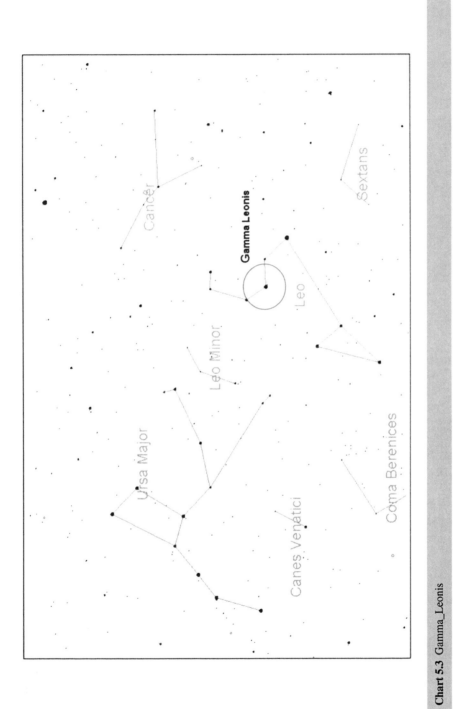

Chart 5.3 Gamma_Leonis

Chart 5.4 Polaris

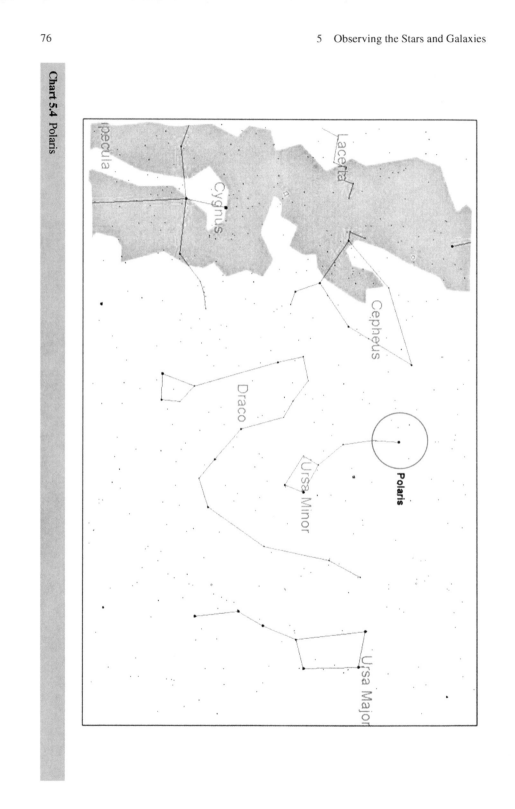

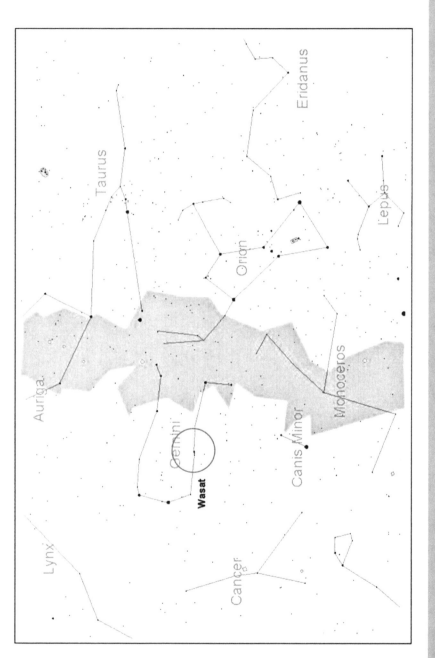

Chart 5.5 Wasat

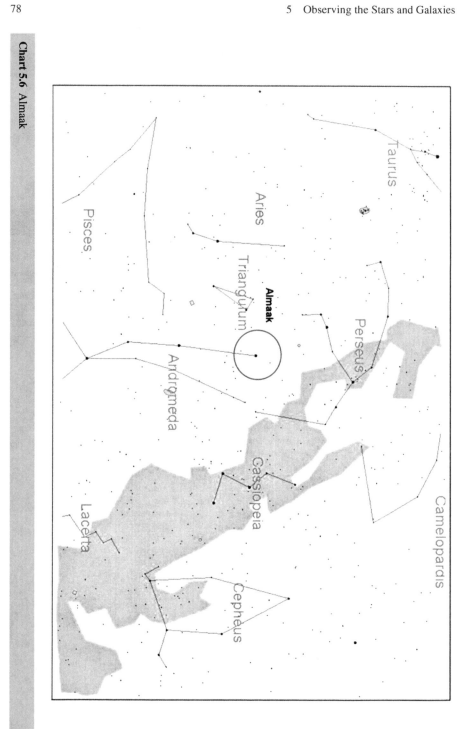

Chart 5.6 Almaak

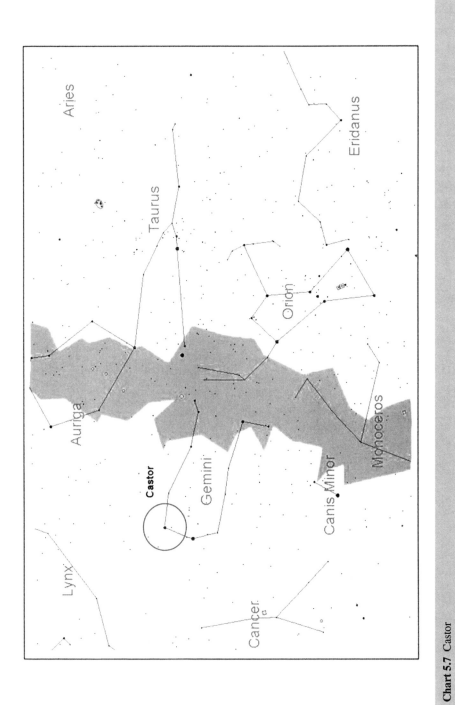

Chart 5.7 Castor

Chart 5.8 Arich

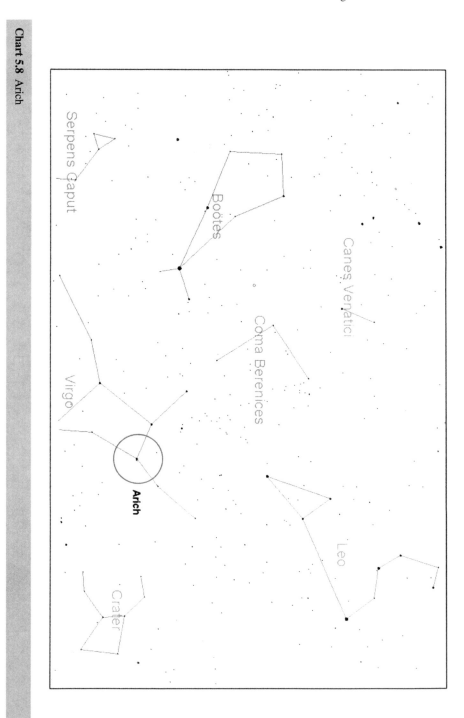

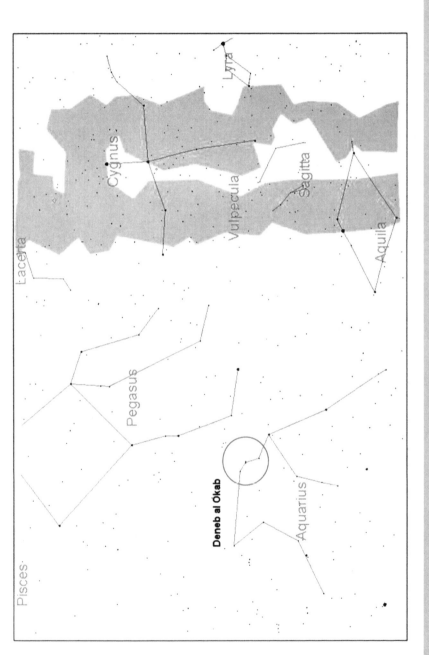

Chart 5.9 Deneb al Okab

Chart 5.10 Al Rischa

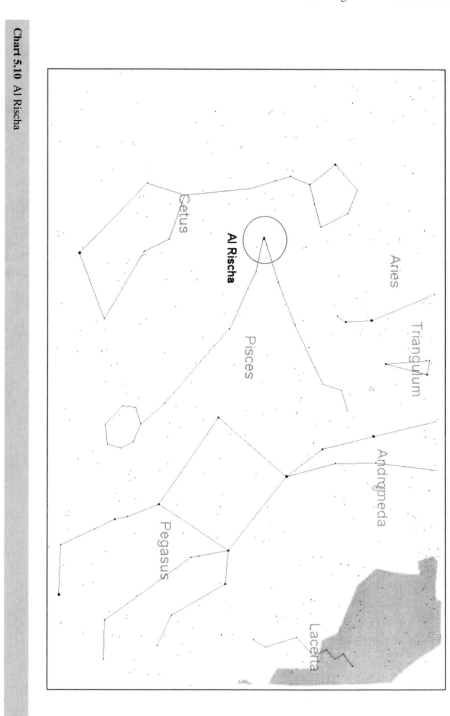

Multiple Stars

We also come across more complicated systems. The classic example is Mizar (Zeta Ursae Majoris) in the Great Bear; it has a naked-eye companion, Alcor, and is itself a very wide, easy double, with unequal components (magnitudes 2.3 and 4.0). Between Mizar and Alcor is a faint star which is 'in the background,' considerably further away and not gravitationally linked to the others. See Chart 5.11.

Close to the brilliant blue Vega. in Lyra, is a naked-eye double. While testing the 110 mm Tasco Luminova reflecting telescope, we looked at Epsilon Lyrae as a test. A top-quality 3-inch telescope will (just) show that each component is again double, so that we have a quadruple system. The Meade ETX 90 (at 3½-inch aperture) showed all four stars clearly, as did the 4½-inch Tasco and the 5-inch Sky Watcher. See Chart 5.12.

In the Orion Nebula, south of the Hunter's Belt, we find 'the Trapezium,' Theta Orionis; all four members can be seen with a 3-inch telescope, though it is best to use a fairly high magnification of at least 100×. See Chart 5.13.

It is always fascinating to range round the sky looking at these double stars. Start with easy pairs, such as Albireo and Mizar, and then graduate to more difficult objects – such as Antares, the red star in the Scorpion, which has a companion 0 magnitude 5.4 which looks decidedly green by contrast. The separation is 2″.7, so that ought to be an easy pair, but in fact it is not, because the secondary is so overpowered by the glare of Antares itself.

Chart 5.11 Mizar

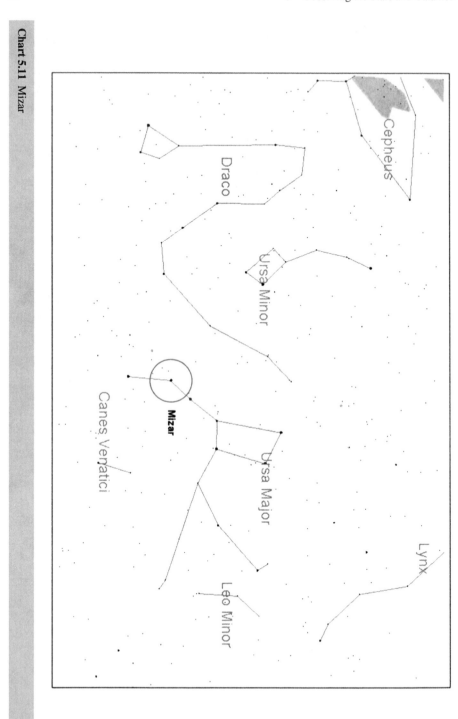

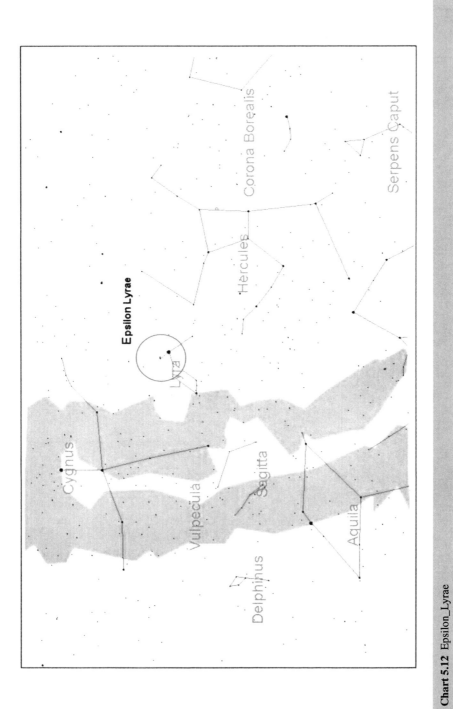

Chart 5.12 Epsilon_Lyrae

Chart 5.13 Trapezium

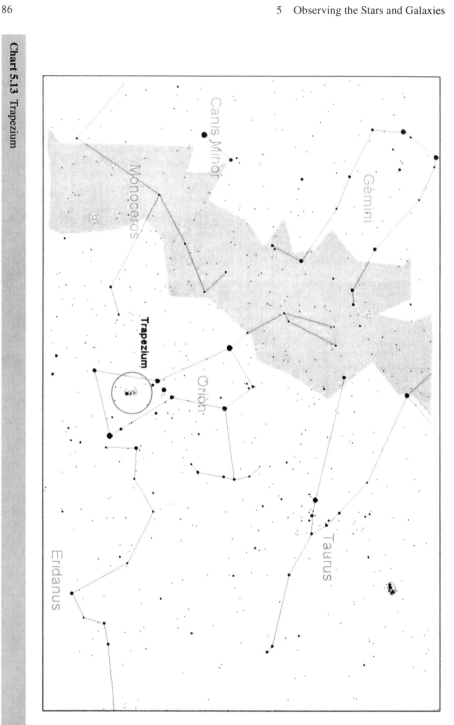

Variable Stars

Most stars shine more or less steadily over long periods, but there are some which do not. These *variable stars* are of various types. Some brighten and fade regularly over periods of from a few days to a few weeks; these are known as Cepheids, after the prototype star, Delta Cephei in the far north of the sky. Mira variables, named after Mira Ceti in the Whale, have longer periods and greater magnitude range; virtually all of them are red (types M, R, N or S). There are 'eruptive' variables, whose behaviour is less predictable, and some which are completely irregular. Different again are the eclipsing stars, such as Algol in Perseus, which is a binary with a period of 2½ days, and one component much brighter than the other. When the secondary passes in front of the primary, Algol fades by more than a magnitude, remaining at minimum for a mere 20 minutes before starting to recover.

Some variables remain bright enough to be followed with the naked eye; Betelgeux in Orion ranges between magnitudes 0.3 and 0.9 in a very rough period of several years, while Delta Cephei varies between magnitudes 3.5 and 4.4 in a period of 5.4 days. There are many variables which never become dim enough to be lost in a small telescope such as can be bought by mail order, but of course a bigger telescope can go down to a fainter magnitude.

Amateurs can do valuable work in studying variable stars; there are so many of them that professional astronomers are grateful for amateur assistance. What has to be done is to estimate the brightness of the variable by comparing it to nearby stars which do not change.

As an illustration, consider R Leonis near Regulus. It is a red Mira variable, with a magnitude range of from 5.4 to 10.5 and a period averaging 313 days – with Mira stars, unlike the Cepheids, the periods and amplitudes are not absolutely constant, and no two cycles are identical.

Locate R Leonis, most easily with a low magnification. In the same field you will see 18 Leonis (magnitude 5.8) and 19 Leonis (6.1). If R is midway between them in brilliance, its magnitude will be 6.1. There are various methods in use for making these estimates – each observer will have his own favorite – but with practice it is possible to make estimates correct to a tenth of a magnitude, which is accurate enough for most purposes. See Chart 5.14.

There are various hazards. For example, it is never too easy to compare a red star with a white one, and many variables are fiery red. Some variables are awkward enough to lie well away from any suitable comparison stars, and many Mira stars – in fact, most – stay within the range of small telescopes only when near maximum. For example, R Andromedae, period 409 days, can just become a naked-eye object at maximum, but at minimum fades down to magnitude 15.

Mira Ceti itself, period 332 days, is a naked-eye object for several weeks in most years, though there are spells when maximum occurs when Mira is above the horizon only during daylight (maxima fall about a month later each year). See Chart 5.15.

The best procedure here is to learn the telescopic position when Mira is bright, so that it can be followed later; at minimum it falls to magnitude 10. Not all maxima are equal; some never rise above magnitude 5, though on one or two occasions in

Chart 5.14 R Leonis

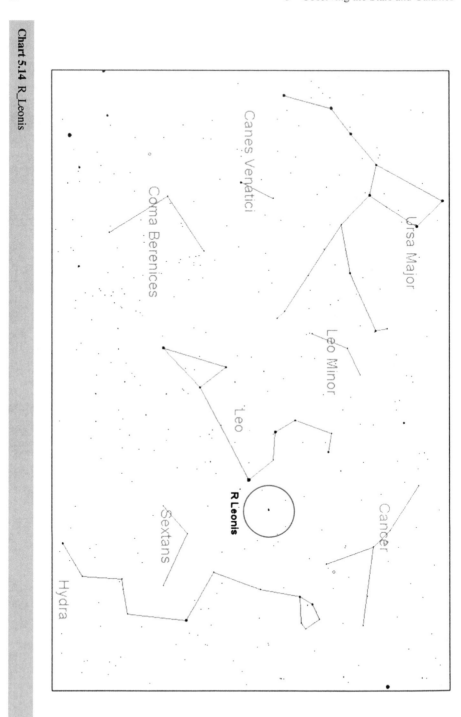

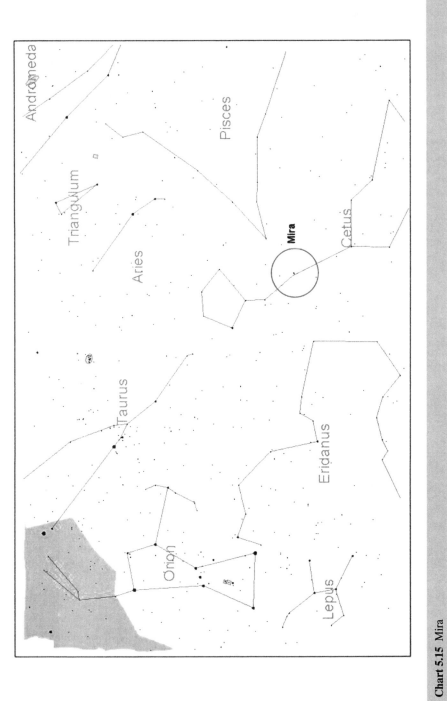

Chart 5.15 Mira

the past Mira has become brighter than the Pole Star. The period, too, can vary for a week or so to either side of the 332-day mean. The only other Mira star which can become fairly prominent with the naked eye is Chi Cygni, in the Swan; range 3.3–14, period 407 days.

One very interesting variable is R Coronae Borealis, in the 'bowl' of the Crown, not far from Arcturus. For most of the time it hovers around magnitude 6, but may unpredictably fall to a very faint magnitude – below 15 – because of clouds of soot which accumulate in the star's atmosphere and mask its bright surface. When the soot is dispersed, R Coronae brightens again. Also in the bowl of the Crown is a star of magnitude 6.8. Look at R with a low power when it is bright, and memorize the field. When, R vanishes below binocular range, you will be able to follow R as long as possible, and keep watch for its reappearance after minimum. See Chart 5.16.

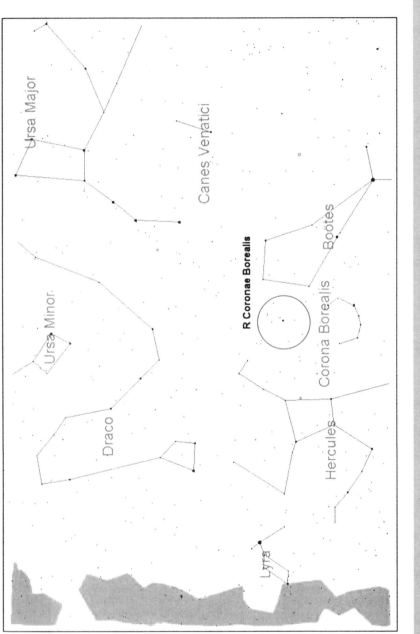

Chart 5.16 R Coronae Borealis

A list of variable stars which remain visible with almost all budget telescopes for part of the time is:

Star	Type	Max	Min	Period	Spectrum
R Andromedae	Mira	5.8	15	409	S
R Aquilae	Mira	5.5	12	284	M
R Cassiopeiae	Mira	4.7	14	430	M
Omioron Ceti	Mira	1.7	10	332	M
R Cygni	Mira	6.1	14	426	M
U Cygni	Mira	5.9	12	462	N
R Geminorum	Mira	6.0	14	370	S
R Hydrae	Mira	4.0	10	390	M
R Leonis	Mira	4.4	11	312	M
R Leporis	Mira	5.5	12	432	N
U Orionis	Mira	4.8	13	372	M
R Serpentis	Mira	5.1	14	356	M
R Trianguii	Mira	5.4	13	266	M
W Orionis	Semi-regular	5.9	7.7	212	N
R Sculptoris	Semi-regular	5.8	7.7	370	N
R Scuti	Eruptive	4.4	8.2	140	G to K
R Coronae Borealis	Irregular	5.8	15	–	F

N and *S* stars are exceptionally red. R Leporis has been nicknamed 'the Crimson Star.' Details for these stars are given in various books and software. In the list above we have not included naked-eye or binocular variables.

Novae, or 'new stars,' may appear at any time. In fact they are not new; what happens is that a formerly very faint star flares up and may become bright for a very brief period of a few days, weeks or months. Amateurs are expert nova-hunters, but this is not a task suited to budget telescopes. When a nova appears, it may of course be followed in the same way as any other variable star.

Star Clusters

Star clusters are among the most spectacular celestial objects that can be seen with an entry-level telescope. The sight of a vast ball of brilliant stars hanging in the field of the telescope is astonishingly beautiful.

In 1781 the French astronomer Charles Messier published a catalog of over 100 star-clusters and nebulae. He did this because he was interested in finding comets, and wanted a ready-reference to objects that he might mistake for comets. He numbered these objects, and the *Messier Catalog*, with these 'M numbers' is still widely used (for example, the Hercules globular is M13).

The NGC or *New General Catalog*, drawn up in the 1880s by J.L.E. Dreyer, is the official source: NGC numbers are universally used.

The *Caldwell Catalog* includes most of the bright nebular objects not listed by Messier, and the C numbers are widely used by visual observers. Almost all the M and C objects are within the range of a 4-inch telescope.

Star clusters are of two types, *open* and *globular*. Open clusters are simply collections of stars which are genuinely associated. Some are populous, with hundreds of members, others are sparse. Globular clusters are vast spherical systems living round the edge of the main galaxy; they may contain more than a million stars, but are so far away (thousands of light-years) that they are not prominent. Only three are clearly visible with the naked eye, the Hercules cluster and the far-southern Omega Centauri and 47 Tucanae.

The best known *open clusters* are the Pleiades and Hyades in Taurus, and Praesepe in Cancer (the Crab); all these are prominent with the naked eye. They are large, and probably best viewed through binoculars, because the small telescope field will show only small areas at a time.

Here are a few examples, beginning with:

M.13 (NGC 6205) the 'Hercules globular cluster.' The brightest globular cluster visible from the mid-Northern latitudes – especially London and New York – it is just visible with the naked eye as a dim patch. Begin with low power, and then use as high a power as possible to resolve it into individual stars. See Chart 5.17.

Here is an image of M13 made with a large telescope, reproduced here by courtesy of Yuugi Kitahara:

Unfortunately you won't see it quite like that in a budget telescope, but even visually and with a small instrument, its appearance is still pretty spectacular:

M.11 The 'Wild Duck' cluster in Scutum (the Shield). Easy to find, near the southernmost stars of Aquila, but with binoculars it is easy to pick out. Our small telescope will show a fan-shaped arrangement, with an 8th magnitude star at the apex. The cluster is 5,500 light-years away. The NGC number is 6705. Use a moderate magnification. See Chart 5.18.

M.41 Open cluster in Canis Major (NGC 2287). It is fairly easy to resolve into stars, and easy to locate, as it lies near the brilliant Sirius. Use a low power. See Chart 5.19.

M.35 (NGC 2168) Open cluster in Gemini, near the Orion-facing side of the constellation. M.35 contains at least two dozen stars brighter than magnitude 9, forming patterns of loops and curls. Use a low or moderate magnification. See Chart 5.20.

M.4 (NGC 6121) Globular cluster in Scorpius. It is easy to locate, 1½° west of the red Antares. Identify it with a low power, and then change to a high magnification to resolve as much of it as you can. See Chart 5.21.

Many open clusters, and a few more globulars, are within our range; search through the Messier and Caldwell catalogs.

Chart 5.17 M13

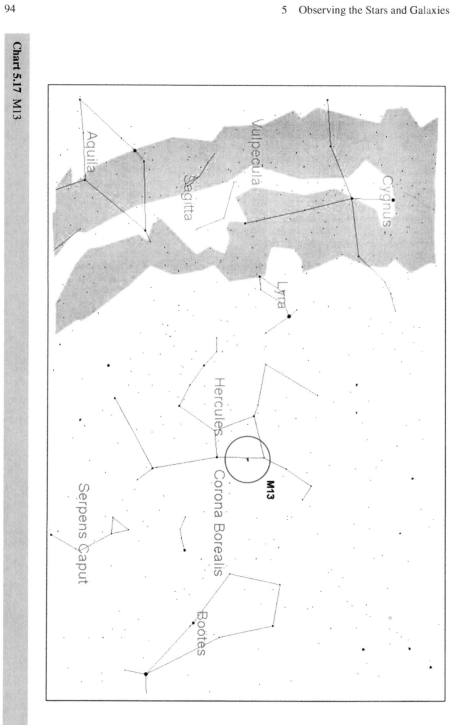

Fig. 5.3 The Hercules globular cluster M13 – the brightest globular cluster visible from northern latitudes

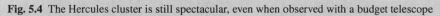

Fig. 5.4 The Hercules cluster is still spectacular, even when observed with a budget telescope

Chart 5.18 M11

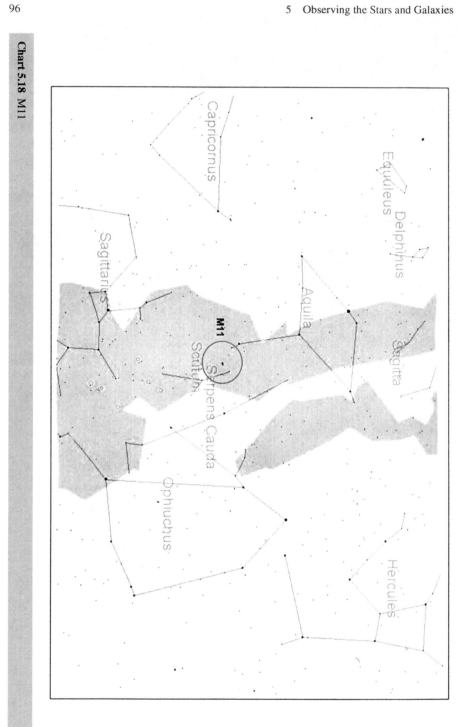

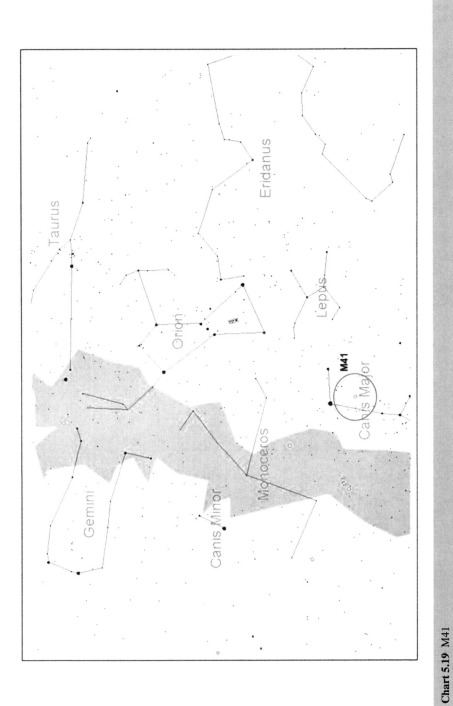

Chart 5.19 M41

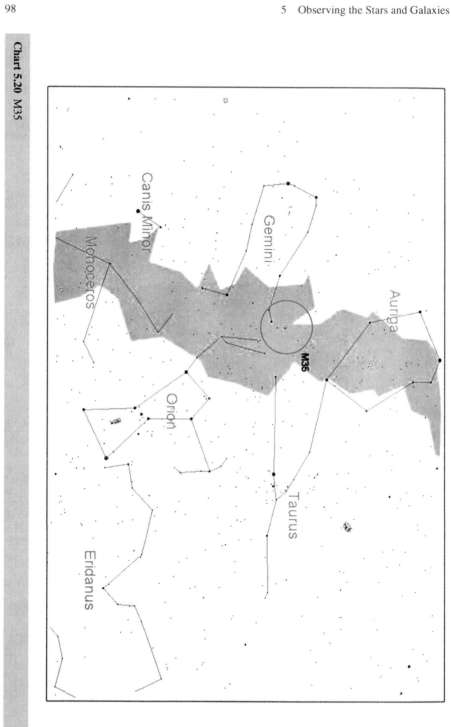

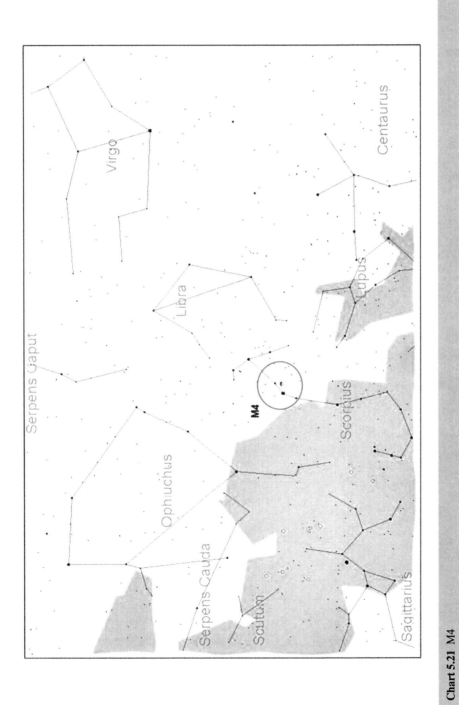

Chart 5.21 M4

Nebulae

Many nebulae are stellar birthplaces – clouds of dust and gas in space, inside which fresh stars are formed. The best-known example is M.42 in the Sword of Orion (not to be confused with the Sword-Handle in Perseus). It is seen as a misty patch with the naked eye, below the belt. See Chart 5.22.

Telescopically M42 is a glorious sight with bright areas and dark rifts; silhou-etted against it is the Trapezium, made up of the four stars of Theta Orionis. It is these stars which excite the nebular material and make it shine. Start with a low power to give a general view, and then change to higher magnification to bring out the finer detail.

Viewed through the Hubble Space Telescope, the Orion Nebula looks like the photograph in Figure 5.5.

In our budget telescope it is far less spectacular – but it is still one of the most impressive sights in the night sky, and by far the most outstanding nebulae. See Figure 5.6.

Averted vision will help you see more of M42 – but you should easily locate the little central group of the four stars of The Trapezium shining through the nebulos-ity. Try higher magnification on the four stars.

M.1, the Crab Nebula in Taurus. This is a very different sort of object – the wreck of a star which was seen to explode as a supernova in 1054, though since it is 6,000 light-years away the outburst actually happened before there were any astronomers on Earth ready to observe it. It is can just be seen near the star Alheka (Zeta Tauri). Finding it with a small telescope is quite a challenge, but well worth doing, though of course no detail can be seen. See Chart 5.23.

M.8, the Lagoon Nebula in Sagittarius (NGC 6523). An open cluster together with a gaseous nebula. It lies near the star 9 Sagittarii, and is an easy object, at first inspection it looks like an ordinary cluster but a closer look reveals the nebulosity. See Chart 5.24.

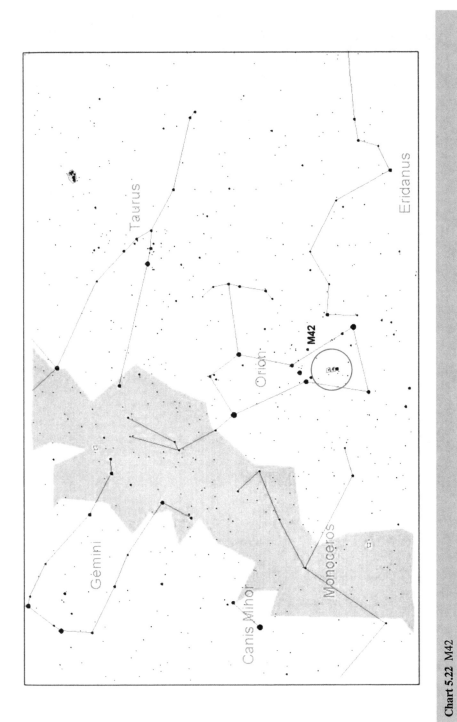

Chart 5.22 M42

Fig. 5.5 The Orion nebula, imaged by the Hubble Space Telescope

Fig. 5.6 Like the Hercules cluster, the Orion nebula is one of the most spectacular (and easily found) deep-sky objects for a budget telescope

Chart 5.23 M1

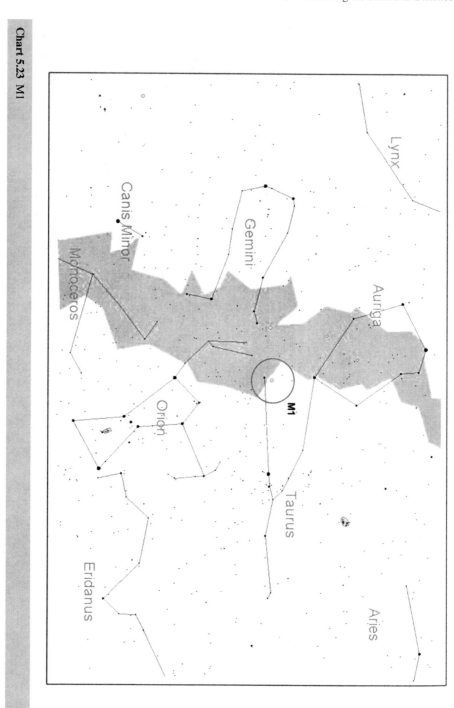

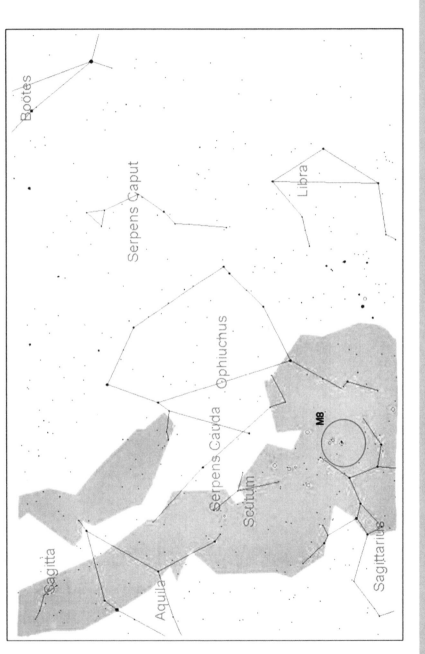

Chart 5.24 M8

Galaxies

Galaxies are external systems, millions of light-years away. The nearest system is M31 in Andromeda, it is just visible with the naked eye, our telescope shows it as a dim oval patch. Use the lowest possible magnification. See Chart 5.25.

It is in fact a galaxy larger than ours, with more than our quota of 100,000 million stars – but it is over two million light-years away. It has to be admitted that in a telescope of modest aperture it is disappointing, looking like a big cotton fluff. It's more impressive when you remember that everything else you can see in the sky with the naked eye is inside our galaxy, and M31 is another great system, broadly similar. Figure 5.7 on page 108 shows a long-exposure image of M31.

But because it is a very dim object, the very best you can hope to see with a small telescope, even on a perfectly clear night, looks more like this:

There are other galaxies in the Messier and Caldwell catalogs, but they are less rewarding in a budget telescope – larger telescopes are needed to bring out their forms. However, you can have fun finding and identifying them. This applies equally to the various clusters and nebulae which are within range of our equipment.

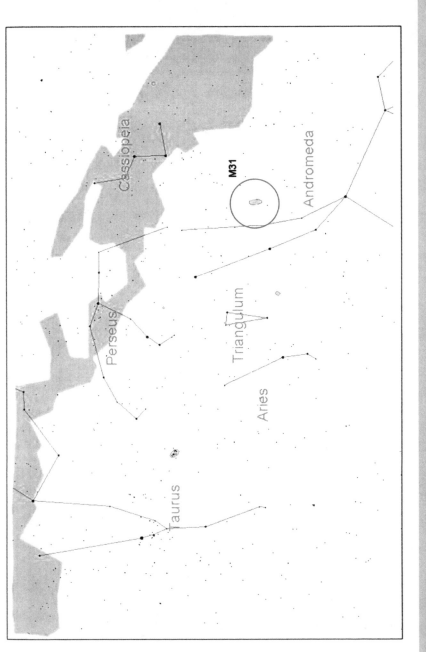

Chart 5.25 M31

Fig. 5.7 The Andromeda galaxy is the most distant object visible with the naked eye – this long-exposure image shows a lot of detail

Fig. 5.8 In a budget telescope, the Andromeda galaxy will appear visually rather dim and fuzzy

Chapter 6

Photography and Imaging

The eye can only see what is in front of it at any given time; no matter how long you stare at a dim object, it won't seem to get any brighter. This is not the case with a camera, which is why photography is such a powerful tool for astronomers. When you use a camera to take a normal photograph the shutter (be it mechanical or electronic) is usually open for a tiny fraction of a second, to 'stop' the motion of moving objects. Typically this ranges from 1/100 second to 1/2,000 second for a modern camera, hand-held. As you might expect, ten times more light arrives on the CCD (charge-coupled device) chip (that replaces the film in a modern digital camera) if the shutter is open for 1/100 second than it does if it is open only for 1/1,000 second. Obviously, this is not the same way that the eye works.

The longer the camera exposure, the more light arrives on the CCD, which accumulates the result. Up to a point, the length of time for which the CCD can accumulate light is unlimited (although the 'point' is generally less than a minute), so for viewing faint astronomical objects, the camera is far superior to the eye.

A camera has another advantage. At low light levels, our eyes lose the ability to see color; it is the results of a trade-off between color vision and sensitivity. Evolution has made our eyes useful under a whole range of conditions, and for obvious reasons the ability to see a faint – and perhaps moving! – shape in the dark has been far more important to us than the ability to see color at night. Cameras can respond to color at almost any light level, so if we photograph astronomical objects we can see what colors they 'really' are.

Digital photography has two disadvantages for astronomy. The first is inherent in the CCD chip. Electronic 'noise,' similar to the effect that you get on your TV screen when the set is mis-tuned, is produced in the CCD by ambient heat. This limits the length of time for which the CCD in a digital camera can accumulate light. After a certain time, the image gets too 'noisy' to use.

P. Moore and J. Watson, *Astronomy with a Budget Telescope: An Introduction to Practical Observing*, Patrick Moore's Practical Astronomy Series, DOI 10.1007/978-1-4614-2161-0_6, © Springer Science+Business Media, LLC 2012

The second disadvantage of digital cameras is that with the exception of the most expensive ones (that's coming on a thousand dollars/pounds sterling) they don't have interchangeable lenses, so you can take the lens off. The ability to remove the camera's lens is, as you'll see, very helpful for astronomy.

Specialized astronomical 'CCD cameras' can be purchased, and we will mention these only briefly because any one of them costs substantially more than any of our budget telescopes! Astronomical CCD cameras work on exactly the same principle as 'normal' digital cameras, and since they lack focusing, flash, storage media and a lens they are actually a lot simpler than even the most basic digital camera. They have just one vital extra – a refrigerated CCD chip! Because the 'noise' that limits the CCD's performance at the long exposure times needed for astronomy is caused by heat, you just have to get rid of the heat…so in an astronomical digital camera the CCD chip is cooled by a solid-state refrigeration system that lowers its temperature down to more than 55°F (that's about 30°C) *below* ambient temperature.

It has to be said that the budget telescopes we looked at are all less than ideal for photography. The main reason is the lack of an electric drive motor that will track the stars smoothly in Right Ascension; without such a tracking motor, stars look like streaks instead of dots because of their motion across the sky. With even slight magnification, it's impossible to work without a motor and use manual controls unless the telescope mounting is incredibly solid. Manual slow-motions sound as if they ought to help, but with any budget telescope they are too coarse for smooth tracking of any object seen through the telescope. And of course you have to avoid the slightest vibration in the mounting.

So we're sorry, but close-up photographs of the stars, deep-sky objects and planets are out. You *can,* however, obtain very impressive photographs of the Moon and Sun – which are bright enough to need only short exposures. Images of meteors and comets are also possible with very simple equipment.

Altazimuth mountings, even if electrically driven (see the review of the Sky-Watcher on page 139), can't be used for photography because of *field rotation*. This means that even if the star you are aiming at is held perfectly in the middle of the field, the rest of the image rotates around it, causing streaking. An electrically driven equatorial mounting is therefore essential if photography is your main aim, Although…

Photographing the Moon

The Moon is the easiest target for beginning photography because it's bright, close, and has plenty of surface detail to record.

Eyepiece-Projection Photography

You can often get good results just by aiming the digital camera down a low-power eyepiece. You will probably need a camera that has the facility to disable the auto-focus, because this can sometimes be counterproductive by wanting to focus on the edge of the eyepiece field instead of the Moon.

If your camera has a zoom lens (most do), begin by setting it to somewhere about mid-range, and get the best result by experiment. Always use the lowest power eyepiece, and align the telescope so that the terminator is in the middle of the field.

Now very carefully aim your camera through the eyepiece from an inch or two away, and gradually move in closer until you can see the Moon's surface on the viewfinder screen. The object is to get as close as possible to the eyepiece without touching it (which would cause vibration). It takes a fair amount of skill to keep the image in the viewfinder; you have to line up the camera perfectly. Once you have done so, is the image sharp? If it looks perfectly sharp, take a picture: if your camera has a 'multiple photo' setting, take a burst of several photos. If the image isn't sharp, then slightly adjust the focus of the eyepiece (not the camera) and try again. Try different zoom settings. Aim for the sharpest possible image.

It isn't entirely easy – and it won't be perfect the first time you try. The picture in Figure 6.1 is the result of John's first attempt with a digital camera and a 4-inch reflector!

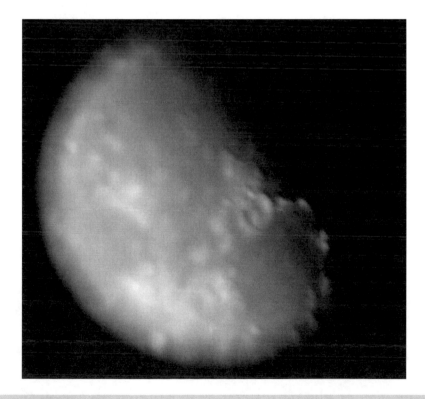

Fig. 6.1 First image of the Moon with a digital camera – not a good start!

But it gets better. After some practice it should be possible to get one out of ten shots that is useful; the hard part is lining up the camera in all five axes and then holding it still during the exposure…This is where a digital camera is supreme – you can take dozens of pictures without worrying about wasting film!

The picture in Figure 6.2 was taken with a Panasonic Lumix TZ7 digital camera and a 5-inch reflector, using just this technique.

Fig. 6.2 This image of the Moon was also taken by aiming a camera down the eyepiece of a 5-inch Newtonian

Camera Mounts

The business of getting the camera lined up with the eyepiece is solved by using a purpose-made camera bracket. These are available from various suppliers (look on the Internet) and generally clamp round the eyepiece barrel, with some kind of bracket to support the camera. See Figure 6.3.

Fig. 6.3 An excellent purpose-made camera-bracket, the oddly named 'Universal Digiscoping'

To avoid vibration, you should use the self-timer on your camera, or a long 'cable release' if there is provision for one. The down-side is that the weight of the camera might need to be counterbalanced, so that the telescope can still be aimed properly. As it isn't usually possible to buy a special counterbalance weight for a budget telescope, you may have to improvise.

Prime-Focus Photography

If you can remove the lens from your camera, you can try *prime-focus photography*. In effect, your telescope becomes a high quality, very long-focus lens for your camera. You need to remove both the camera lens and the telescope eyepiece, and fix the camera so that the telescope's objective forms its image on the film.

It is possible to buy an adapter that enables your camera to replace, in effect, the eyepiece. Such adapters are available to fit many DSLRs and connect them to a tube that fits a standard 1¼-inch eyepiece, or to some other part of the telescope tube. You can buy them from astronomical equipment suppliers, who advertise in the astronomy magazines.

It is worth mentioning that DSLRs tend to be *heavy*. Not as heavy as the old film SLRs (which were usually made of metal and not polycarbonate), but heavy. There is therefore a chance that a DSLR might actually be too heavy to fix to a small telescope safely. Regular digital cameras are usually much lighter.

With a camera mounted at the prime focus of the telescope (and almost always with some form of counterweight) you can focus an image of the Moon sharply. The image of the Moon at the prime focus will be smaller than the image obtained with eyepiece projection. Typically you should be able to get the full diameter of the Moon – about ½ degree – into the frame. See Figure 6.4.

Fig. 6.4 A prime-focus image of the Moon

Start by using the automatic exposure setting, but experiment with exposure times. Most digital cameras have an option to increase or decrease the exposure from what the automatic system decides.

Photographing the Sun

Remember that observing the Sun is potentially dangerous, and you must never look at the Sun through any sort of optical aid.

The best way is to fit a full-aperture solar filter (as described above) and use exactly the same procedure as you would use for observing the Moon. The image in Figure 6.5 of a group of small sunspots was taken through a 3½-inch reflector, using a Fujicolor FX1500 digital camera.

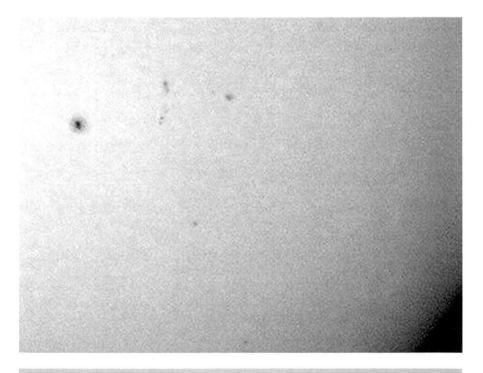

Fig. 6.5 Sunspots, taken in exactly the same way as the Moon image in Figure 2.6, but of course with a full-aperture solar filter fitted to the telescope

Photographing the Deep Sky

The deep sky includes everything outside the Solar System. Without an electric R.A. drive, magnified photographs of deep-sky objects are out of the question, but it is certainly possible to take moderately long-exposure photographs of star fields and extended objects such as M42 (the Orion Nebula).

Piggyback Photography

The minimum requirement is an equatorial mounting with slow-motion drives. The idea is to fix your camera to the telescope, use the equatorial mounting to track the movement of the stars across the sky, and at the same time use the telescope itself as a high-powered 'guide telescope' to keep the camera pointed as exactly as possible.

Some of the better budget telescopes actually have a purpose-made bracket for attaching a 'piggyback' camera to the telescope tube, although not all of the manufacturers mention it in their specifications. You may need a *tripod bush* (this is probably not supplied) that passes through the bracket to attach the camera – they are obtainable at most camera shops. If your telescope isn't equipped with a mounting bracket, it usually isn't too difficult to rig up something suitable. The object is to end up with the camera rigidly mounted on the telescope tube.

Most digital cameras don't have a mechanism for holding the shutter open indefinitely – the equivalent of the 'B' (brief time) setting on film cameras. But some – and I can quote the Panasonic Lumix TZ7 here – have quite long time exposure settings, plus something rather clever to improve long-exposure photographs. You may remember that we mentioned that a major limiting factor with CCD cameras (as opposed to film) is thermal noise generated in the chip. The Panasonic Lumix TZ7 (and its successor the TZ10) have, among a plethora of special settings, something called 'Starry Sky'. This gives the option of 15, 30 or 60 seconds exposure time. But better still, after the exposure it automatically makes a 'dark frame' – that is a second exposure of the same duration, but with the shutter closed. Then it subtracts the dark frame from the 'real' exposure (clever, these computers!) to arrive at the difference between the two. That's basically just the stars…

Start off by using a wide angle, and progress to a longer zoom as your ability to track the sky gets better. Choose a clear night, when the stars are shining brightly and there is no full Moon to out-shine them. Carefully align the equatorial mounting (as described above) and aim at a suitable patch of sky. Using the highest-power eyepiece (even if this produces a fuzzy image), aim the telescope at a bright star somewhere in the field you want to photograph, and use the slow motion controls to move it to the center of the field.

Begin with a practice run…well, several practice runs. Use the R.A. slow-motion to hold the star in the exact center of the high-power eyepiece. It isn't as easy as it sounds. The trick is to move the control very slowly and evenly without making the telescope vibrate. Keeping the flexible drive shaft of the R.A. control bent through at least 45° helps reduce the amount of vibration transmitted to the mount. If despite your careful tracking the star moves out of the field within a minute or so, requiring correction by means of the declination control, you should re-align the equatorial mounting.

Once you are satisfied that you can keep the star more or less centered for 3 or 4 minutes, you are ready to try taking a photograph.

Get your guide star in the middle of the field and open the shutter...using the camera's delay for 5 or 10 seconds is best, to give vibrations in the telescope time to die down.

The exposure of Orion in Figure 6.6 was made using the Lumix TZ7, at 60 seconds, set to 'Starry Sky.' There's still a certain amount of blurring of the stars caused by less-than-perfect tracking, but it's much better than the equivalent shot with the camera mounted on a tripod, that showed severe 'trailing'.

Fig. 6.6 A 60 second exposure of the constellation of Orion, using a digital camera 'piggy-backed' onto a manually driven equatorially mounted telescope

If you are fortunate enough to be around when a major comet appears in the sky, then piggyback photography with a budget-priced telescope equipped with slow-motion drives is quite possible. Comets move quite quickly against the starry background and so even a telescope with a motor drive needs to be continuously manually corrected. Comets at their closest approach can appear as quite large objects and so lend themselves to being photographed this way – magnification is not needed! This photograph of comet Hyakutaki, see Figure 6.7, was taken by Grant Privett:

Fig. 6.7 Comet Hyakutaki, photographed in less then optimum conditions!

This is an un-retouched photograph, and it suffers somewhat from underexposure and the glow on the horizon caused by city lights about ten miles away. In the next Chapter you can see how this image can be enhanced using a PC.

It is possible to get quite impressive results from piggyback photography, but manual tracking tends to be gruelling without a motor drive. This image of M13 was taken a few years ago, using a film SLR equipped with a 135 mm long focus lens, manually tracking for about 20 minutes at f/3.5, piggybacked onto a 4-inch reflector used as a guide scope. Guiding was very difficult indeed because of the longer focal length camera lens, and once again the star images aren't exactly round, due to tracking errors. The picture was copied to a CD-ROM by the processing house and then slightly enhanced using the brightness and contrast adjusting features of Paint Shop Pro...which of course leads neatly on to the next chapter.

Fig. 6.8 The Hercules cluster, imaged with a 'piggybacked' film camera; probably about the limit of most people's endurance for manual tracking!

Chapter 7

Image Processing Software

Digital cameras all use the .JPG format as standard, which is ideal for subsequent processing on a PC.

For users of budget telescopes and regular digital cameras it is probably best to begin work with general-purpose software, partly because of the limitations of the original images, partly because of the cost, and partly because such software can be used for all sorts of other fun things as well. The various specialized astronomical image-processing packages are meant to be used with astronomical CCD cameras. Paradoxically, the simplicity of CCD cameras makes their pictures easier to process, because the software has access to a 'raw' image.

Apart from your computer and the relevant software, you will need a photo-quality inkjet printer – they are very inexpensive these days – and a supply of glossy photo paper to print on.

As an example of software, here (Figure 7.1) is Paint Shop Pro, produced by JASC Sofware Inc. It is an excellent 'middle of the road' application with many very useful features but a price that makes it affordable for most.

P. Moore and J. Watson, *Astronomy with a Budget Telescope: An Introduction to Practical Observing*, Patrick Moore's Practical Astronomy Series, DOI 10.1007/978-1-4614-2161-0_7, © Springer Science+Business Media, LLC 2012

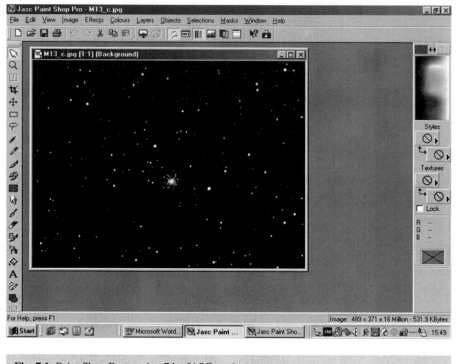

Fig. 7.1 Paint Shop Pro version 7 by JASC: main screen

Note the vast array of tools and options scattered around the edge of the image – these make Paint Shop Pro a wonderfully flexible tool for all sorts of graphic work. For astronomical images, the main features you will want to use for your astronomical photographs affect the image brightness, contrast, color, sharpness, and something called 'unsharp masking.' All these are controllable to a considerable extent.

Just how much image processing constitutes 'enhancement' and how much is 'art' rather depends on who you are asking, and on what you want the picture for. A photograph intended for a picture-frame on the wall can be processed as much as you like to make a good-looking image. If you want an image for more scientific purposes then you need to be more restrained. Done carefully, image processing can bring out hidden details and improve an astronomical photograph immensely, but done badly it can result in loss of fine gradation, loss of faint stars, and at worst an artificial-looking and obviously 'over-processed' image.

There is really no substitute for practice.

For technical reasons, images that were digital in the first place are more easily processed than images that have been transferred from photographic negatives. This is because a 'scan' of a negative can never provide more information than exists on the film whereas there may be initially invisible data in a digital image that can be brought out by processing.

Sharpness

Using the 'sharpen' or even the 'sharpen more' feature will usually improve an astronomical image. However, if you overdo it, the background noise (grain) will be sharpened to a point where it detracts from the image, and spurious detail will appear.

Figure 7.2 shows an original image (made with a small SCT, so North is up and East is to the left):

Fig. 7.2 Basic Moon image – the inset shows the area around Maurolycus, enlarged three times

The insert shows the area around Maurolycus, enlarged three times to show more clearly the effects of processing best: here it is (Figure 7.3) after sharpening.

Fig. 7.3 After enhancement using the 'sharpen' tool

...but Figure 7.4 shows how it looks after *too much* of the sharpening process:

Fig. 7.4 Over-enhancement – the effect of too much sharpening

Contrast and Brightness

Enhancing the contrast can make an image look much more striking, but it is easy
to lose details such as faint stars. The most useful feature is the ability to use judi-
cious control of contrast and brightness to darken a sky that is either over-exposed
or afflicted with light-pollution. Increasing contrast but also increasing the brightness seems to be best. Figure 7.5 shows a photograph of the constellation of Aquila,
reproduced by courtesy of Grant Privett:

Fig. 7.5 Aquila, photographed through a severely light-polluted sky

...and here it is (below, next page) after processing, which mainly involved increasing the contrast, reducing the brightness, and 'de-vignetting' (that is, correcting for the fact that inexpensive camera lenses – especially zooms – tend to give a brighter image in the centre than they do round the edges). De-vignetting is done by making a deliberately very blurred version of the original – so blurred that astronomical objects are fuzzed into invisibility – reversing it (i.e. making it into a negative), and merging it with the original, unblurred image. This is quite an advanced technique, but all you need besides skill is a PC and software, and as you can see, the results are startling. See Figure 7.6 on page 127.

Grant Privett's photograph of comet Hyakutaki was taken (pre-digital!) with a 70–200 mm f/3.5 zoom lens on an old Praktika SLR, 3 minutes exposure on ISO 400 color film. Because Grant didn't have access to a film scanner at the time, he scanned a print. Figure 7.7 on page 127 shows the original.

Fig. 7.6 The same photograph of Aquila, after enhancement work

Fig. 7.7 Grant Privett's photograph of Hyakutaki

Figure 7.8 shows how it looks after some serious enhancement work.

Fig. 7.8 Hyakutaki, after enhancement

The processing involved increasing the contrast, reducing the brightness, and getting rid of the worst of the red cast (which was exacerbated by the contrast increase). This is done by adjusting the hue, and also by reducing the color saturation of the affected parts of the image. The picture was also sharpened slightly.

It is important to understand that you can process the image a lot or a little. The original photograph of this comet shows a rather beautiful sky which has been removed during processing – but of course you can stop at any point. Just be sure to keep the original image files (save the new ones under different file names) in case you mess it up and have to start again!

Unsharp Masking

This is a very powerful technique that can bring out a wealth of hidden detail, especially from digital images. The odd name is derived from the original photographic technique, which involved the use of a cut-out opaque mask that was used during the enlarging process. Unsharp masking is a method for enhancing sharpness and detail in faint objects, especially in stars. It works with pictures that are not too grainy (it also enhances the grain!) and can make big improvements. Overdone, it results in dark circles round stars and a strange artificial look, or worse it will introduce detail and even extra 'stars' where none exist. It can also mess up the colors. Figure 7.9 shows a photograph of a star field.

Fig. 7.9 Original image of a star field

Fig. 7.10 The same star field after enhancement with the 'unsharp mask' tool

Figure 7.10 above shows it after moderate processing.

The improvements are obvious and genuine. A range of variables within the unsharp masking routine (radius, strength and clipping) that allow you to make changes to this feature – and Paint Shop Pro (among others, of course) allows you a preview to see what's going to happen. Once again, lots of practice is essential. Figure 7.11 is a terrible warning about what can happen if you overdo unsharp masking:

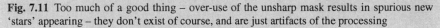

Fig. 7.11 Too much of a good thing – over-use of the unsharp mask results in spurious new 'stars' appearing – they don't exist of course, and are just artifacts of the processing

None of these 'extra stars' exist! They are no more than thermal effects and artifacts of the camera's JPG image compression algorithm, turned into visible dots by too much 'enhancement'.

If you want to find out more – much more – about image enhancement, then you could do no better than get a copy of Grant Privett's book, *Creating and Enhancing Digital Astro Images: A Guide for Practical Astronomers*, published by Springer.

Color Saturation and Hue

Faint colors in astronomical photographs can be enhanced by increasing the color saturation. Don't overdo it, or the images will look ridiculous. The Sun when photographed through a full-aperture solar filter will usually appear an odd color, as a result of the filtering. Reducing the color saturation usually helps make solar images appear more 'natural' by making them whiter.

Orange skies caused by poor street lighting (sky glare from a city is generally orange, from sodium lamps) can be reduced by using the hue control to make the sky blacker – but be aware that it will also affect astronomical objects, which will become correspondingly bluer as you shift the hue away from red.

Wherever you live (almost) there will be cloudy nights – now that's when you get your opportunity to set about processing your last batch of images!

Chapter 8

Trying Out Two Typical Budget Telescopes

Finally, it seemed appropriate to devote some space to providing at least a couple of reports on one or two of the better budget telescopes.

For obvious reasons (we both live here!) the telescopes we tried out were purchased in the UK. Similar (or identical) models are available in the US, but it's worth pointing out that consumer goods, including telescopes, are in general somewhat less expensive in America than in Europe, so a direct currency conversion into dollars will probably yield a price that is on the high side.

We actually tried out numerous low-cost telescopes as a prelude to writing this book, often (but not always) with quite gratifying results. During the course of the tests we used the little 90 mm Meade ETX – one of which both authors happen to own – as a 'control' for comparing the other telescopes. Although the ETX is small, and by the standards of top amateur telescopes inexpensive, it costs more than twice as much as the most expensive of the telescopes we looked at. It does however come from the Meade Instrument Corporation of California, arguably the major manufacturer of small telescopes. It is representative of the highest standards of optical excellence, and is mechanically sound.

P. Moore and J. Watson, *Astronomy with a Budget Telescope: An Introduction to Practical Observing*, Patrick Moore's Practical Astronomy Series, DOI 10.1007/978-1-4614-2161-0_8, © Springer Science+Business Media, LLC 2012

Tasco Luminova 675× Reflector Telescope

Fig. 8.1 Patrick with the Tasco Luminova

John ended up with the job of unpacking, assembling and getting this one working!

Unpacking and Assembly

The Luminova came in a brown cardboard outer box, containing a polythene bag, containing a nicely color-printed cardboard box, containing a white cardboard box, containing several white cardboard boxes each printed with their contents. In short, it was very well packed!

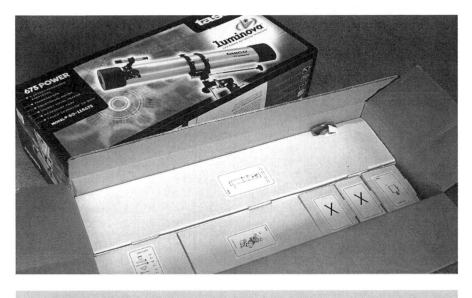

Fig. 8.2 The Luminova, as shipped

Inside the boxes are the telescope, equatorial mounting, finder scope, three eye-pieces (25 mm, 12.5 mm and 4 mm), a 3× Barlow lens and a small toolkit – which even includes a nice little screwdriver. There is also a CD-ROM, intriguingly labeled 'Tasco Skywatch' but with no clue at all as to what it's for! More about this later.

The 'Owner's Manual' wasn't very good at all. Containing only seven pages of information, it seeks to describe – with the aid of two pages of diagrams not a great deal clearer than cave paintings – the assembly of no less than five different models of Tasco telescope (three reflectors and two refractors). That said, assembling the telescope was quite easy. It took about half an hour. However, I wonder how long it would have taken someone unfamiliar with astronomical telescopes to do it... a lot longer, I suspect!

The tripod legs (they are made of metal) bolt to the bearing head, and a triangular spacer (that doubles as an eyepiece holder) is held in place by three wing-nuts. Three big knobs screw into the side of the tripod legs to allow for height adjustment.

The telescope tube is secured to the bearing head by heavyweight metal clamping rings that allow it to be removed very easily by releasing two knurled knobs. A substantial steel rod screws into the bearing head to support the (surprisingly heavy) counterweight.

Two flexible shafts with knobs on the end, for declination and R.A. slow-motion, slip on to their shafts and are secured by knurled knobs. (Don't assemble it like a mail-order catalog photograph, with the declination knob at the bottom of the telescope, as far away from the eyepiece as possible!)

Finally, you are told to set the polar axis angle to your angle of latitude and do up the clamp. It wasn't immediately clear why the whole thing flopped down again, until a more careful examination of the drawings in the manual revealed a screw (which I had ignored) that both holds the latitude adjustment in place and allows fine control of its setting.

There is a finder scope, pre-assembled into a plastic mounting that simply bolts to the side of the telescope tube.

The equatorial mounting, once assembled, is surprisingly solid, and the slow-motions work smoothly. All in all, I think this telescope is well-made for the price, and there are several nice details – the inclusion (as ever, unmentioned in the manual) of a camera attachment screw, to allow for 'piggyback' photography, for example.

Alignment

I lined up the viewfinder with the main telescope in daylight, as I usually do. This job took quite a long time but was no more nor less difficult than any other 'three screw' system. The plastic mounting of the finder isn't really stout enough and once aligned you need to be careful not to bump the guide scope and knock it out of alignment.

Collimation

There are three big securing bolts and what are presumably collimation screws on the back of the metal cell containing the primary mirror. Similarly, there are the usual three screws on the front of the secondary mirror support.

Fig. 8.3 The Luminova secondary mirror is supported by a single stout strut

The Owner's Manual makes no reference to collimation. Reflecting telescopes need their mirrors precisely aligned. It isn't hard to do but if the alignment – known as collimation – is wrong, the image seen will be distorted and stars may look odd shapes – stars that look like little fish is a common sign of bad collimation.

Fortunately, when I checked the telescope optics, the collimation of the telescope as taken out of the box proved to be perfect, for all intents and purposes.

First Light

Patrick says: 'I first turned the 110 mm Luminova on Mizar, using its lowest power. The image looked very good, but when I compared it with the image produced by the 90 mm ETX the stars appeared less defined and marginally less focused. This is to be expected, as the Schmidt-Maksutov design of the ETX has a reputation for optical excellence and the spherical mirror of the Luminova and its relatively inexpensive eyepiece could not be expected to be as good. But it was actually quite a creditable performance.

Albireo showed its contrasting colors nicely, although not as well as it did when we used the little Meade.

'I thought I would next try a fairly severe test of resolution and turned the Luminova on Epsilon Lyrae, the famous 'double double'. You can find it using Chart 5.12, on page 85 of this book. The star consists of four components. At low magnification there appear to be two stars, just like many other wide doubles, but with increasing power you can see that each of the two is itself a binary star. The close pairs are similar in brightness and are separated by only 2.9″ (seconds of arc) and 2.3″ respectively.'

The ETX will just about split the close pair, on any reasonable night.

Dawes' Limit sets a theoretical maximum resolution for a telescope. In the case of the 90 mm ETX it is about 1.28″, and for the 110 mm Luminova it should be 1.05″.

The Luminova did rather well. It split the close binaries at least as easily as the ETX, and the eyepieces supplied allowed me to use higher magnification than I could get with the ETX, which helped even further.

M31, the Andromeda Galaxy – a spiral galaxy, magnitude 4.8 – looked bright and crisp, and rather to my surprise (although I'm not sure why I was surprised, considering the difference in aperture), visibly brighter than it did in the Meade.

The lowest-power eyepiece supplied with the Luminova – 25 mm – provided a much narrower field of view than the ETX 'Super Plossl' eyepiece of about the same focal length. This was a disappointment only until I remembered the relative costs!

That said, the Luminova has a standard eyepiece diameter so if you want to buy better eyepieces for it, you can...

I also liked the mounting, which was much better than expected with 45× and 112×, though the Barlow strained it.

Lunar and planetary detail was good – a belt on Saturn was visible right away in average seeing. With the slow motions, stability was very reasonable. Commendable.

Tasco Skywatch CD-ROM

On to the mysterious CD-ROM. We put it in the drive of John's PC as instructed, and the CD auto-started and asked us if we wanted to install Tasco Skywatch. How could we possibly know? We had no idea what it was for. Still, in a spirit of experiment we pressed on and told it yes. After a while it asked us if we wanted to install 'Quicktime'.[1] In for a penny, in for a pound, we thought – why not? We told it yes again. These various installations went smoothly.

Now for the exciting part: John clicked on the Skywatch icon that had appeared on the Windows Desktop... and we can now reveal that Tasco Skywatch is an example of what is known as a 'planetarium' program: that is, it shows the night sky from any location and for any time of day.

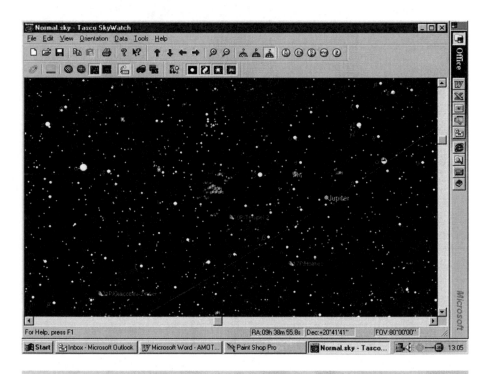

Fig. 8.4 Tasco Skywatch main screen

---

[1] *Quicktime* is a freeware program, supplied by Apple Inc., that allows a PC to play many different formats of moving pictures and sound. A lot of PCs have it pre-installed. Regular upgrades are provided automatically by Apple.

We used it for about an hour, at the end of which we both came to following conclusion: *Tasco Skywatch* is superb – and far better than any of the 'freeware' we were able to find on the Internet!

If bought separately, we would have expected to pay over £50 for it (that's probably $50 because of price differentials). It's awash with functions. There are all the usual 'find' facilities, a big database of stars and non-stellar objects, a 3D solar system plotter, chart, map printing, orientation, even photographic images of many interesting clusters, galaxies and other objects. All in all it seems to do everything you could possibly want of such an application – it even has a 'dark-adaption' button that turns the display red, so you can use it outside in the dark (on a laptop) without spoiling your night vision.

The only puzzle is, why on earth don't Tasco and their retailers make more of it?

It's certainly a big selling point and adds a lot of value for anyone who owns a PC – that's most amateur astronomers these day – but only if you tell people about it! *Tasco Skywatch* deserves more of a showing in the catalog.

The Sky-Watcher Explorer 130P SupaTrak™ Telescope[2]

As they used to say in Monty Python, now for something completely different.

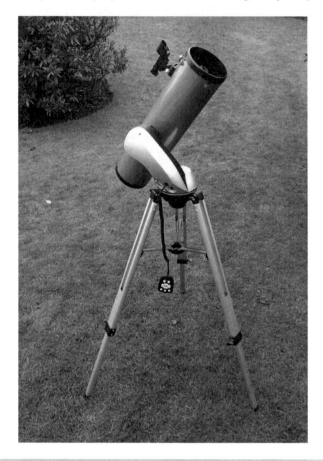

Fig. 8.5 The Sky-Watcher Explorer 130P SupaTrak™ Telescope – possibly as good value as it's name is long

Unpacking

The telescope, tripod, and all its other components arrived well-packed in a brightly colored box. There was a label on the outside, saying that the solar filter shown on the box (it looks as if it's made from mylar film) isn't available in the UK, Scandinavia or the Baltic states. Presumably the safety standards are higher in these countries!

---

[2] This instrument was kindly provided on loan by Broadhurst Clarkson & Fuller Ltd., of Telescope House, Starborough Farm, Marsh Green, Kent TN8 5RB, UK. www.telescopehouse.com.

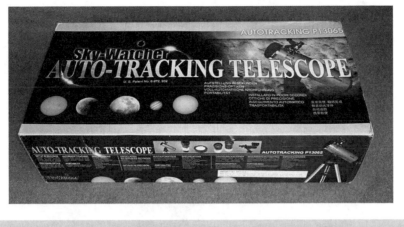

Fig. 8.6 The Sky-Watcher, as shipped

The SupaTrak Mounting

This 5-inch (130 mm) telescope has a completely standard Newtonian optical design, but features an altazimuth mounting with electric drives for automatic tracking. At the time of writing (Spring 2011) this kind of mounting is very unusual in a budget telescope – although the (much more expensive) computer-controlled 'go-to' mountings use the same basic idea.

Fig. 8.7 The SupaTrak™ is a single-arm altazimuth mounting

The Sky-Watcher Explorer 130P's altazimuth mounting is electrically driven on both axes, and a small computer automatically adjusts the speeds of the motors so that the telescope can track any object in the sky. To set it up, you first need to tell it the latitude of your observing site, which takes only a couple of minutes. This has to be done only once because the computer remembers it even after the power is disconnected.

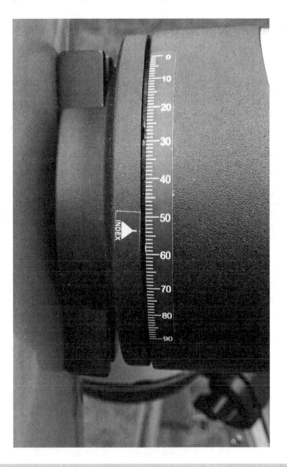

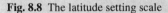

Fig. 8.8 The latitude setting scale

To prepare the Sky-Watcher Explorer 130P for use, you lock the mounting onto its sturdy tripod (there is a single enormous knob to tighten it), fit the telescope tube to the mounting (another single knob to lock it in place), and slide the 'red dot' finder (see Chapter 2) into its bracket. Setting up the automatic tracking is just a matter of moving the telescope to the zero mark on the altitude scale and pointing it towards the north. They even supply you with a little magnetic compass! That's it.

You then use the hand controller to aim the telescope, using the red dot finder, at your target. Press the buttons to select auto-guide, and there you are.

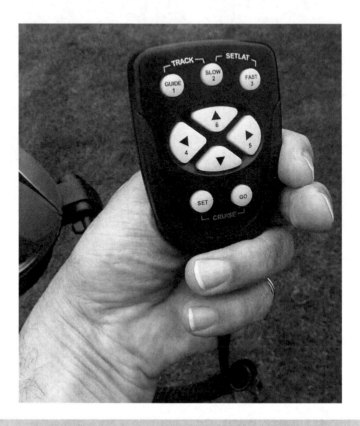

Fig. 8.9 Sky-watcher hand controller

Does it work?

Well, the answer is an unequivocal 'yes!' The automatic tracking works very well indeed. We tried it out on a clear evening in the Spring and it kept our target – M42 in Orion – firmly in the middle of the lower-power eyepiece field indefinitely. Even using the highest magnification (the 10 mm eyepiece plus the 2× Barlow lens, which gives 130×) the image drift was very slow. Tracking would probably have been almost perfect if we had taken more care with the northerly alignment when we set the telescope up.

One final piece of praise for this unusual mounting: the operation of the slow-motion buttons that are used to inch the position of the image into the middle of the eyepiece field or to correct tracking if necessary, is superb. The buttons are laid out properly – that is, the left button moves the image in the eyepiece left, the top one moves it up. More important – and anyone who has struggled with one of the far more expensive computer-controlled telescopes from major American manufacturers will really appreciate this – pressing the slow-motion buttons has *immediate* effect, and with no time-lag or overshoot.

The Telescope

The Sky-Watcher Explorer 130P telescope is a straightforward Newtonian design. It has a metal tube, with four very thin tensioned blades to hold the Newtonian flat mirror. They can be adjusted by four milled knobs. (Compare the picture of the *Tasco Luminova* on page 136 which uses a single stout secondary support).

Fig. 8.10 The Sky-watcher secondary mirror is supported by four very thin tensioned steel vanes

The Skywatcher Explorer 130P comes with two eyepieces: 25 mm and 10 mm, as well as the 2× Barlow lens. There is a magnetic compass for finding north and a little black plastic handbag to keep the unit's batteries in. And a tiny screwdriver, for some reason.

Collimation

The telescope was correctly collimated when it arrived, so we didn't need to alter anything. The primary mirror can be tilted by adjusting three recessed screws at the bottom of the tube (Hah – maybe *that's* what the screwdriver's for!) but the secondary mirror seems to require a small hex key (that's an allen key in the UK). There are no instructions for collimating the telescope, but really it's best left alone unless you already know how to do it.

Alignment

Aligning the red-dot finder with the telescope is very easy. Once you have the telescope pointed at a bright star (and tracking it!) two control knobs are used to move the red dot around to coincide with the star's position in the sky.

First Light

As mentioned above, Orion was our first target. The low-power 25 mm eyepiece provided a view of M42 at least as good as that shown in Figure 5.6 on page 103 (find on Chart 5.22 on page 101) and the higher power of the 10 mm eyepiece split Theta Orionis (the Trapezium) easily into its four components. Star images were sharp right to the edge of the field, visible as points in the low-power field and with no tell-tale signs of under or overcorrected optics.

The rack and pinion focusing was commendably smooth.

Epsilon Lyrae (we tried out the Tasco Luminova on it) not being visible at this time of year, we turned the Sky-Watcher Explorer 130P towards Alpha Geminorum (Castor, see Chart 5.7 on page 79). At 65× magnification, using the 10 mm eyepiece, the two components of this bright star (magnitudes 2 and 3, about 6″ separation) could just be seen as separate stars. With the Barlow lens as well – that's 130× magnification, about the practical maximum for a 5-inch primary mirror – the dark space was visible between the two components. Zeta Ursae Majoris (Mizar), see Chart 5.11 on page 84, is an easier bright binary, with a separation of 14″. Its brightness makes it an interesting test for small telescopes: the Skywatcher Explorer 130P did very well and even at its highest power showed the two components without flare. Those thin secondary supports ensured that only the faintest of diffraction spikes were visible. Excellent.

Because this telescope seemed so good, we compared it with John's Meade EXT-125 (a 5-inch Maksutov-Cassegrain design of about the same aperture), rather than with the 3½-inch EXT-90 we used as a standard for the other budget telescopes we've looked at.

Yes, of course the ETX-125 is optically better. It's a given that a Maksutov is better corrected than a Newtonian, and the Meade's much longer equivalent focal length – 1,250 mm compared to the Sky-Watcher's 650 mm – meant that to get about 65× magnification we could use the 25 mm eyepiece instead of the 10 mm one, which helped the eye relief and the apparent field of view. But make no mistake, both of the Sky-Watcher Explorer 130P eyepieces are very good indeed, for budget lenses.

Overall, this was the best small telescope we tested for this book, and it represents excellent value. The optical system is very good, and all the bits and pieces you need are supplied as standard. But that 'SupaTrak' mounting is quite outstanding. It's definitely the best so far for visual observing, unless you want to spend substantially more on a full 'go-to' system. And even then, the SupaTrak slow-motion controls are probably better!

Chapter 9

The Next Steps

Telescopes

The picture in Figure 9.1 shows John's 'portable' telescope, a 10-inch (250 mm) Meade LX200, shown with a piggybacked CCD camera. The laptop PC can operate the *Starlight Xpress* MX5-C CCD camera and – simultaneously – automatically track celestial objects to a very high degree of accuracy. Unless you are a weight-lifter, this is about the largest telescope that can sensibly be regarded as remotely portable for one person.

P. Moore and J. Watson, *Astronomy with a Budget Telescope: An Introduction to Practical Observing*, Patrick Moore's Practical Astronomy Series, DOI 10.1007/978-1-4614-2161-0_9,

Fig. 9.1 John's 10-inch (250 mm) Meade LX200. But is it 'portable'?

Apart from his Meade ETX and solar telescope, all Patrick's telescopes have permanent observatories in Selsey. Figure 9.2 shows his 12½-inch Newtonian, in a 'double run-off' protective building.

Fig. 9.2 Patrick's home-built 12½-inch Newtonian has a double run-off protective building

Budget telescopes will give surprisingly good results, but of course they are limited, and before long the real astronomical enthusiast will start to cast around for something capable of providing more power. This is often where trouble begins, because telescopes of larger aperture are either good or cheap – almost never both – and in fact may not be the answer for everyone.

There was a time, not so long ago, when amateurs favored making their own reflectors (not refractors; lens-grinding was always too much of a problem for most people). It is true that making a mirror of, say 8–12 inches aperture is more laborious than really difficult, but it is immensely time-consuming, and the beginner must be prepared for many failures before producing his first reasonable mirror. Mounting, of course, is fairly straightforward, the main essential being to make it firm. A good rule here is to work out the maximum weight of your proposed mounting – and then multiply by three.

Nowadays it is usual to buy a telescope complete, so where to start?

It might not necessarily be a good idea to go for the biggest, most expensive telescope you can afford. It all depends on your circumstance. Patrick is lucky enough to have space for permanent observatories, which he found was the best answer for him. John is the first to admit that he uses his 5-inch ETX125 far more often than the LX200. Why? The answer is about available time. It takes over an hour to get the 10-inch LX200 in place, aligned, and ready to go. The ETX125 can be set up and running in around ten minutes.

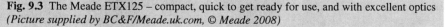

Fig. 9.3 The Meade ETX125 – compact, quick to get ready for use, and with excellent optics *(Picture supplied by BC&F/Meade.uk.com, © Meade 2008)*

The smaller telescope is compact, and also much, much lighter than its big brother. It is normally used with its internal batteries and can be packed into a single (admittedly rather large) carrying case, so can even be taken on holiday – unless you are travelling by air, of course.

So try not to contract 'aperture fever' and buy bigger and bigger telescopes, regardless. Remember that what you go for should depend on (a) your main interests, (b) the amount of money you propose to spend, and (c) the practicality of using the instrument.

For most people, a reflector is probably the answer, because a reflector is much cheaper than a refractor of equal power. However, always be on your guard. There are some poor-quality reflectors on the market, and they do not always betray themselves at first inspection. If you choose a telescope from a reputable manufacturer such as Meade or Celestron, and from a reputable supplier, you will be safe enough. Look at the bullet points at the end of Chapter 2, and especially be wary of any second-hand telescope until you have been able to test it!

Visual Deep-Sky Observers

If you want to look at faint objects, but don't usually want to image them, a *Dobsonian* mounted Newtonian reflector is clearly the cheapest way to start with a large-aperture telescope.

Named after an American monk, the Dobsonian features an altazimuth mounting designed to be moved very easily and smoothly by hand. A Dobsonian cannot easily be motor-driven without very complicated computers or a fairly uncommon accessory known as an *equatorial platform*. That said, a large Dobsonian – 12-inches upwards – will satisfy the deep-sky enthusiast who wants a wide field, low magnification, and the greatest possible light-grasp for his money, but who is not interested in photography. Aperture for aperture, Dobsonians are by far the best value.

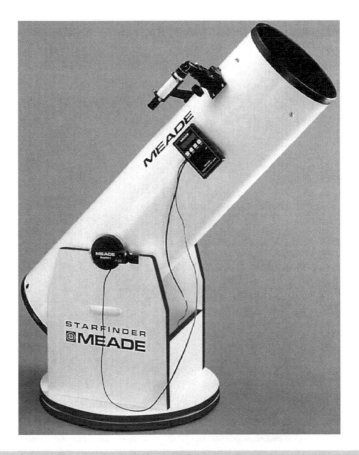

Fig. 9.4 A Dobsonian telescope. 'Dobs' represent the best big-aperture value for visual deep-sky observers *(This picture shows the Meade 'Starfinder', and was supplied by BC&F/Meade. uk.com, © Meade 2002)*

Planetary, Imagers, and General Observers

A good 6-inch Newtonian on an equatorial mounting, with drive, can be obtained for about $300 (£300 in the UK). Aperture for aperture, they are probably the best value for an equatorially mounted telescope. Yet on the whole, a catadioptric telescope is probably the best 'step up' after your budget starter.

As mentioned above, both authors of this book own and use 3½-inch (90 mm) Meade EXT Maksutov-Cassegrain catadioptric telescopes, which cost about $500 in the US and about £500 in the UK. These telescopes are optically superb, portable, and are equipped with a good drive. You can buy a safe solar filter. You can lift them up with one hand. The Meade *Autostar*™ hand controller contains a computer system that allows the telescope to find objects for you. Set it up, press the right buttons, look through the eyepiece, and there – behold! – is the Crab Nebula.

Fig. 9.5 The Meade ETX90, used as a standard of excellence for the budget telescopes we tested *(Picture supplied by BC&F/Meade.uk.com, © Meade 2008)*

Meade is not the only company to make computer-controlled telescopes, of course – far from it. This computer-controlled telescope is Celestron's 'Nexstar' 5-inch (125 mm) Schmidt-Cassegrain catadioptric telescope.

Fig. 9.6 Another excellent American telescope – the Celestron 'Nexstar' 5-inch (125 mm) Schmidt-Cassegrain *(Photograph reproduced here by courtesy of Celestron International, www.celestron.com)*

Where does it end? Well, if you have about $14,000 (or pounds) to spend, you could get a 16-inch Meade LX200 ACF...

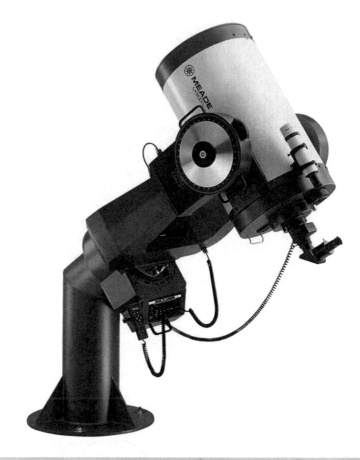

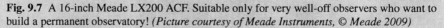

Fig. 9.7 A 16-inch Meade LX200 ACF. Suitable only for very well-off observers who want to build a permanent observatory! *(Picture courtesy of Meade Instruments, © Meade 2009)*

... but of course there are those other matters to be considered, like the views from your chosen location and the cost of building a permanent observatory around it!

Solar Observers

Finally, there is a lot to be said for observing the Sun. It is our closest star and presents an ever-changing subject for astronomical observation. (I suppose we should say, except at the time of a sunspot minimum when the Sun can look about as interesting as a cue ball for a couple of years.)

If you are fascinated by observing the Sun, a 4-inch (100 mm) with a full-aperture solar filter is adequate, and is portable, though the solar observer will find this less essential than the night worker; light pollution is less important and you can observe from anywhere there is sunshine. For serious solar work a telescope with a hydrogen-alpha filter is recommended; Patrick's is shown here.

Fig. 9.8 Patrick with his hydrogen-alpha solar telescope

A solar telescope working in the hydrogen-alpha band will give endless enjoyment, and you can watch the prominences and other phenomena in safety. The cost is unfortunately high – the least expensive is the Coronado PST (see Chapter 4) which costs about $600, and you would need to pay at least $3,000 (or £2,500) for the telescope in the photograph – but of course it has a limitless lifetime provided that it is well treated.

Even Bigger and Better

1. This is no space to go into detail about still more powerful telescopes – and that's not what this book is about – but you can refer to specialized books.

Imaging

If you have become fascinated with photographing the heavens, then the mounting will be even more important than the telescope. The major American manufacturers offer superb value for money and generally excellent quality. An equatorial mounting is *essential* for photography (a computer-controlled altazimuth won't do because the field in the eyepiece rotates as the telescope tracks). Electric slow-motion controls are commonplace and pretty well essential as well. A 'Go-to' computer that will find and track invisibly faint objects is enormously useful. It is worth bearing in mind that all Meade altazimuth-mounted telescopes have the facility for the mounting to be tipped over for equatorial use (although the larger models require an accessory 'super-wedge'). The Autostar has software settings for both modes, to enable the telescope to 'go-to' any object whichever way you have set the mounting up. This means you can use the telescope for imaging the sky… a wonderful area of astronomy, but unfortunately not something we have room for in this book.

Mike Weasner's book, *Using the Meade ETX: 100 Objects You Can Really See with the Mighty ETX* (Springer, 2002) contains a lot of useful information about this range of telescopes, and for a selection of more specialized astronomical imaging books, check out *www.Springer.com/sky*.

A simple digital camera, web-cam, DSLR camera, or purpose-made astronomical CCD camera (in increasing order of cost) all provide excellent results if used properly. CCD cameras in particular are superb for obtaining good images under light-polluted skies.

Given all this technology, and even given unlimited money, astrophotography is never easy. It requires dedication, hours and hours of work, and not the least an understanding and sympathetic partner…

The Internet

Bigger and better telescopes are not the whole story. Astronomy is very much a collective activity, and the advice to 'join a local society' is good. However, that may not always be possible for everyone, so equally good advice is 'get on the Internet'. If you can do both, that's ideal!

The worldwide community of amateur astronomers makes tremendously good use of the Internet. There are, quite literally, *thousands* of web sites belonging to astronomical societies and even individuals, and between them they contain vast amounts of interesting information. Use one of the search engines, like *Google* or *Yahoo!* or *Bing* to find the sites you want. Search on something basic and relevant like 'Amateur Astronomy' and you will come up with at least several million hits. You can try to refine your search by adding more qualifications – not always successfully. For example, 'amateur Moon images' came up with 12,400,000 hits when we tried it! 'Amateur crater images' made 164,000 hits.

Most universities and also astronomical equipment manufacturers also have their own sites, and there is a lot of free ('freeware') or inexpensive ('shareware') astronomical software that you can download and use. Among the most interesting sites are the Forums and Message Boards, which allow people to debate any and every issue. NASA and JPL sites provide the latest news, along with spectacular state-of-the-art images.

Books and Specializing

The Internet provides an incredibly rich resource for amateur astronomers, but it is totally unstructured. If you want to learn more about astronomy, and in particular if you want to learn about a specific branch of astronomy, a book is still the best place to start.

Yes, as authors, of course we have a vested interest in books, but if you think about it, they are unbelievable value. For less than the cost of a meal out for one or – for our English readers – for the cost of an ordinary return rail fare between London and Basingstoke, you can buy a good book that will provide hours of engrossing reading, and possibly a source of reference for years to come.

If you want to develop your interest in astronomy, it's unlikely that you will want to remain a generalist. Most of us develop an interest in one specific branch of the science, and it is here that a book can help you make the right choices. Some of the best can be found at *www.Springer.com/sky*.

Whichever branch of astronomy you choose, we wish you good luck, and once again, 'Clear skies'!

Photographs: Acknowledgments

The 'professional' photographs of astronomical objects are by kind permission of NASA and the National Space Science Data Center, unless otherwise credited in the text.

All other photographs are by the authors, unless otherwise credited in the text.

The 'budget telescope views' began with the authors' images – unless otherwise credited in the text – and were processed using *Paint Shop Pro version 7* to make them look as similar as possible to the actual visual impression, as seen through the eyepiece of a 3–4-inch telescope of average-to-good quality. Getting an accurate match to what we could see in the eyepiece proved to be much more of a challenge than we thought it would be, but the end result of several iterations is actually quite realistic.

The 'finder' charts were all adapted from charts produced using *Starry Night Pro*, by Sienna Software Incorporated, of Ontario, Canada.

P. Moore and J. Watson, *Astronomy with a Budget Telescope: An Introduction to Practical Observing*, Patrick Moore's Practical Astronomy Series, DOI 10.1007/978-1-4614-2161-0, © Springer Science+Business Media, LLC 2012

Index

163